Urban Agriculture and
City Sustainability II

WITPRESS

WIT Press publishes leading books in Science and Technology.
Visit our website for the current list of titles.
www.witpress.com

WITeLibrary

Home of the Transactions of the Wessex Institute.
Papers contained in this volume are archived in the WIT eLibrary in volume 243 of WIT
Transactions on Ecology and the Environment (ISSN 1743-3541).
The WIT eLibrary provides the international scientific community with immediate and
permanent access to individual papers presented at WIT conferences.
Visit the WIT eLibrary at www.witpress.com.

SECOND INTERNATIONAL CONFERENCE ON
URBAN AGRICULTURE AND CITY SUSTAINABILITY

Urban Agriculture 2020

CONFERENCE CHAIRMEN

Stefano Mambretti
Polytechnic of Milan, Italy
Member of the WIT Board of Directors

Jose Luis Miralles i Garcia
Polytechnic University of Valencia, Spain

HONORARY CHAIRMAN

Francisco Jose Mora Mas
Polytechnic University of Valencia, Spain

INTERNATIONAL SCIENTIFIC ADVISORY COMMITTEE

Organised by
Wessex Institute, UK
Polytechnic University of Valencia, Spain

Sponsored by
WIT Transactions on Ecology and the Environment

WIT Transactions

WIT Transactions

Wessex Institute
Ashurst Lodge, Ashurst
Southampton SO40 7AA, UK

Editorial Board

Urban Agriculture and City Sustainability II

Editors

Stefano Mambretti
Polytechnic of Milan, Italy
Member of the WIT Board of Directors

Jose Luis Miralles i Garcia
Polytechnic University of Valencia, Spain

WITPRESS Southampton, Boston

Editors:

Stefano Mambretti
Polytechnic of Milan, Italy
Member of the WIT Board of Directors

Jose Luis Miralles i Garcia
Polytechnic University of Valencia, Spain

Published by

WIT Press
Ashurst Lodge, Ashurst, Southampton, SO40 7AA, UK
Tel: 44 (0) 238 029 3223; Fax: 44 (0) 238 029 2853
E-Mail: witpress@witpress.com
http://www.witpress.com

For USA, Canada and Mexico

Computational Mechanics International Inc
25 Bridge Street, Billerica, MA 01821, USA
Tel: 978 667 5841; Fax: 978 667 7582
E-Mail: infousa@witpress.com
http://www.witpress.com

British Library Cataloguing-in-Publication Data

A Catalogue record for this book is available
from the British Library

ISBN: 978-1-78466-381-0
eISBN: 978-1-78466-382-7
ISSN: 1746-448X (print)
ISSN: 1743-3541 (on-line)

The texts of the papers in this volume were set individually by the authors or under their supervision. Only minor corrections to the text may have been carried out by the publisher.

Preface

This volume contains papers presented at the 2nd International Conference on Urban Agriculture and City Sustainability scheduled in Valencia but held on line due to the Coronavirus pandemic. The Conference has been organised by the Wessex Institute, UK and the Universitat Politècnica de València, Spain. The Conference was sponsored by the WIT Transactions on Ecology and the Environment.

The challenge of the sustainable city is more important for our society every day and every year. Every year the population in the cities grows in the world. Food production and food sovereignty are essentials for sustainability in the cities. We are living at a time when researchers from all over the world are looking for new forms of food production in cities.

The scale of modern food production has created and exacerbated many vulnerabilities and the feeding of cities is now infinitely more complex. As such, the food system cannot be considered secure, ethical or sustainable. In the last few years, there has been a rapid expansion in initiatives and projects exploring innovative methods and processes for sustainable food production. The majority of these projects are focused on providing alternative models that shift the power back from the global food system to communities and farmers improving social cohesion, health and wellbeing. It is therefore not surprising that more people are looking towards urban farming initiatives as a potential solution.

These initiatives have demonstrated that urban agriculture has the potential to transform our living environment towards ecologically sustainable and healthy cities. Urban agriculture can also contribute to energy, natural resources, land and water savings, ecological diversity and urban management cost reductions.

Moreover, urban agriculture becomes progressively another way of understanding the city. Thereby the city is transformed by innovative rewilding processes. The impact urban agriculture can have on the shape and form of our cities has never been fully addressed. How cities embed these new approaches and initiatives, as part of new urban developments and a city regeneration strategy is critical.

In fact, urban agriculture includes a lot of multidisciplinary approaches to a new perspective of the city like: urban metabolism, green infrastructure, productive urban landscape, peri-urban agriculture, city food chains, market gardens from development and planning point of view; organic farming, patchwork farms, roof top farms, green roofs, vertical greening and farms, hydro-aqua-aeroponics systems, green houses or food networks from urban farms point of view; community cohesion and job creation, community gardens, co-producers or education and training from socio-economics point of view; and transport and distribution, waste food recovery, healthy cities, care farming, biofuel production, street quality, eco-cities or air pollution reduction from innovation and benefits. The development of appropriate urban agricultures systems is one of the essential concepts to achieve sustainable cities and a way to improve the food sovereignty.

Urban agriculture is a research field that represents a great challenge but also a great opportunity to improve the green infrastructure and the production of healthy food in our cities and their surroundings.

Papers from this Conference, like others presented at Wessex Institute conferences, are referenced by CrossRef and appear regularly in suitable reviews, publications and databases, including referencing and abstracting services. They are also archived online in the WIT eLibrary (http://www.witpress.com/elibrary) where they are permanently available in Open Access format to the international scientific community.

The Editors are grateful to the members of the International Scientific Advisory Committee and other colleagues who helped to review the papers, as well as all authors for their significant contributions.

The Editors, 2020

Contents

Section 4: Urban metabolism

SECTION 1
FOOD SOVEREIGNTY

NEW PHASE IN THE PROCESS OF MAINTENANCE OF THE URBAN AGRICULTURAL AREA OF "L'HORTA DE VALÈNCIA", SPAIN

JOSÉ L. MIRALLES I GARCIA
Department of Urban Planning, Polytechnic University of Valencia, Spain

ABSTRACT

On 12 March 2018, the Valencian official bulletin published the Law of "L'Horta de València". In addition, on 20 December 2018, the same bulletin published approval of the Territorial Action Plan for Planning and Revitalizing "L'Horta de València". Both initiatives establish a framework for maintaining the periurban agricultural area of Valencia and allow actions to promote agricultural activities. In 2019, "L'Horta de València" was declared a Globally Important Agricultural Heritage System (GIAHS) by Food Agriculture Organization (FAO) of the United Nations (UN). The municipality of Valencia became the world capital for sustainable food, to promote the Milan Urban Food Policy Pact. Therefore, two years after protection, it is time to review if the new law and planning have produced positive changes in agricultural activities and landscape, and assess which are the current challenges now. Specifically, it's time to know the current situation, problems, limitations and opportunities. This article provides the results of a SWOT analysis about the current situation of agricultural activities and landscape of "L'Horta de València", carried out by semi-structured interviews with a panel of experts involved in elaboration of these documents or who are intervening in management of this space.

Keywords: *agricultural land management, agriculture, Globally Important Agricultural Heritage System, peri-urban management, regional planning, sustainable farming, urban agriculture, Valencia, vegetable farming.*

1 INTRODUCTION

Valencia's "Huerta" (vegetable farm) is an extraordinary peri-urban agricultural space that is the object of many studies. This long-studied space is today a protected space, managed by a special system regulated by law. All studies have in common the positive evaluation of this singular landscape.

According to Antolín Tomás et al. [1], a long time ago in about 1990, the lands of high agrological capacity in the Valencian community represented only 3.9% of the surface of the Valencian territory, in two main spaces: the fertile plain of the Turia river in Valencia, or Valencia's "Huerta"; and the fertile plain of the lower valley of the Segura river south of Alacant. As a referent, according to Boira [2], in 2006, the urbanized areas were 5.41% of the Valencian territory. That is, the surface of high-capacity agricultural lands is less than the surface of artificial lands. Fig. 1 shows the location of periurban agricultural land protected around Valencia today.

In the period of transition to democracy, about 1978, was born a social movement to protect and maintain this singular space [3]. Progressively, a social majority was generated in favor of this idea; however, at the same time, the Valencian community underwent an accelerated process of urban expansion that transformed many rural areas, including an important part of the peri-urban agricultural land of "L'Horta".

Urban expansion on the Mediterranean coast of Spain began in the 1960s [4]. Public administration promoted a big touristic resort in L'Albufera Lake, such as the project of La Manga del Mar Menor, in Murcia, Spain [5]–[8]. This project generated a very strong social movement against the destruction of spaces with environmental value, like L'Albufera Lake, the periurban agricultural area of "L'Horta" and the old Turia riverbed [9].

Many studies analyzed different characteristics of this singular and complex space. The situation and evolution of "L'Horta de València" is a subject for continuous analysis. It is an

Figure 1:   Geographical area of the peri-urban agricultural space of Valencia's "Horta" and zoning for different kinds of agricultural areas, according to their quality [19].

identity element of Valencian society; many authors focus their attention on it. Romero and Francés [10] coordinated their work with an update on global diagnosis. Temes and Moya [11], [12] studied the evolution and dynamics of agricultural landscape changes. Temes-Cordovez et al. [13] researched the evolution of different periurban agricultural areas in Spain. Miralles i García [14] studied the environmental management problem in periurban areas. Marqués Pérez et al. [15] focused their research on evaluating the agricultural system. Romero and Melo [16]; and more recently, Melo [17] have studied the current situation and dynamics of this landscape.

In addition, the Valencia Council organized the "2017 Milan Urban Food Policy Pact Annual Gathering and Mayor's Summit". One of the results was the document "Estrategia Agroalimentaria. València 2025" (Agro-alimentary strategy for Valencia 2025) [18].

As a result of the social movement and studies, in 2006 the "Pla d'Acció Territorial d'Ordenació i Dinamització de L'Horta de València", (Territorial action plan to regulate and revitalize L'Horta de Valencia, in Catalan) finally began to be elaborated upon by the Regional Government that was approved in December 2018 (12 years later). In addition, according to results of analyses of studies begun about the situation of agricultural land of L'Horta, its problems and possible actions to maintain this anthropic landscape; in 2015, the new Regional Government began the elaboration of a law to manage this singular space. Finally, on 6 March 2018, the L'Horta Law [20], or Llei de L'Horta de València, in Catalan, or Ley de la Huerta de Valencia, in Spanish, was passed and published on 12 March 2018.

These two tools fundamentally change the situation, to protect and manage the agricultural land of "L'Horta de València". Note that this land is a natural resource, transformed and managed by farmers. It is only possible to maintain this space or landscape if the agricultural activity is maintained. Obviously, to maintain agricultural activity in this big area, we need a public policy with public–private collaboration, to promote the many activities and initiatives that farmers cannot do in an individual way.

These two tools, the Territorial Plan and Law, have different objectives. The Plan is a passive way to establish planning uses in the territory. The Law establishes a framework to do initiatives in favor of agricultural or complementary activities. That is, it is a tool for creating an active way to maintain agriculture activities. Both are complementary.

The Plan establishes a framework of land use in three general areas [19], within an extended area with a surface 63,400 ha. This surface includes, totally or partially, the territory of 44 municipalities. This area includes the agricultural land of "L'Horta de València" plus a buffer zone around it. Inside this area, there are: the strict area of "L'Horta de València" including urban areas and irrigation infrastructures, with a total surface of 22,900 ha and 40 municipalities. Inside this strict area there are about 11,393 ha of protected agricultural land (named "protected not developable land" in Valencia, according to Urban Law). This is the agricultural land at the initial time of Plan preparation. The Plan includes written and graphic regulations that are mandatory (for all public and private agents, except Central Administration, because regional rules/plans cannot bind the highest level of administration in Spain). The new way of planning and managing protected agricultural land was studied in detail by the collective work directed by Marzal Raga [21].

The Law establishes a complex system to manage this singular territory, based on three axes in addition to "Territorial Action Plan to Regulate and Revitalize the Huerta of Valencia" axes [22]: the elements that make up the "Huerta", the L'Horta Council and the Agrarian Development Plan.

The Law establishes the main management objectives and a check of possible tools. The "L'Horta Council" is an office with a budget that depends on regional administration and participation of the main private agents interested in agricultural activities and environmental values of the Huerta. In fact, the "L'Horta Council" is the managing entity meant to promote initiatives in order to revitalize agricultural activities or others, according to the framework established in the law. Maybe one of its main functions is to elaborate on the Agrarian Development Plan as a tool to improve agricultural activity and its profitability [23].

All these new tools establish a very innovative new framework to manage this singular space and natural resource. For this reason, it is necessary to rethink the new situation, analyzing the possibilities that the new situation allows, and checking whether the objectives are being met.

Finally, in 2019, the United Nations (UN) Food and Agriculture Organization (FAO) registered "L'Horta de València" as a Globally Important Agriculture Heritage System. The report with the proposal is the last international award regarding "L'Horta de València" and its irrigation system [24]. Note that this irrigation system is historical, with a specific historical court to solve problems about irrigation between the farmers inscribed in 2009, as an Intangible Cultural Heritage site by UNESCO [25].

2 OBJECTIVES AND METHODS

As seen in Section 1, the plan and the law about "L'Horta de València" are two tools that have generated a new situation through possibilities and management of the periurban agriculture of Valencia.

In addition, it is necessary to remember some international trends about management of territory; particularly, it is interesting remember two trends that focus on possible new points of view. First, the concept of "smart territory" as an evolution from the "smart city" concept [26]. In fact, it is possible to have an agricultural production system based on demand in real time, by a semi-automatized way. The use of applications via the internet can change the way production occurs, independently of the kind of production (certified organic, not certified organic or conventional). Secondly, the planning of metropolitan areas implies specific criteria for maintaining green infrastructures inside them, and to avoid territory fragmentation [27] as a strategy to transform the territories in sustainable areas.

The aim of this research is to identify the characteristics of the new situation of the space called "L'Horta de València" after the introduction of two new tools, the plan and the law, which are being applied, and also to identify any new challenges.

To achieve this goal, we did a SWOT analysis (strengths, weaknesses, opportunities, threats) as a synthesis of a semi-structured survey, carried out with a group of five specialists. These specialists are people, not politicians, but with a very significant role in the processes to elaborate and design the plan and/or the law to protect and promote "L'Horta" of Valencia. The interviewees work in different sectors interested in maintaining and promoting agricultural activity of this peri-urban space: university specialists, representative agrarian union people that usually do conventional agriculture, farmers that produce (or want to produce organic agriculture but without the interest to certificate it, organic farmer entrepreneurs, and NGO environmental associations. All these people are specialist technicians or professionals in the subject under study. The interviews have been done individually, in part before confinement, because of the Coronavirus situation; and in part after confinement. Confinement has affected the demand and production of agricultural products.

3 RESULTS

Tables 1–4 show a synthesis of the SWOT analysis. The analytical elements were grouped by topic. The elements identified by two (*) or more (**) interviewees have been marked.

As you can see, in the weaknesses, Table 1 presents the largest list of items that includes eight topics. The first weakness detected focuses on governance problems. There is a general awareness that the action of the regional public administration is necessary to implement actions that farmers cannot carry out by themselves. In fact, a great effort has been made by the regional public administration to prepare and approve the Plan and the "L'Horta" Law; but all those interviewees surveyed consider that, at this time, there is a certain lack of initiative to implement the planned actions. Note that this agriculture is by smallholding, so it is very difficult to do coordinated actions based on private initiatives, especially for strategic actions.

Table 1: Synthesis of weaknesses.

Weaknesses	
Governance	Lack of initiative from the public administration (the Horta Council).**
	Horta Law is a social pact to protect and promote L'Horta, but there are pending problems about several urban expansions.*
	Horta Law has not avoided infrastructures such as motorways or railways promoted by the Central Administration, within the protected area.
	Lack of coordination between different social and public agents to achieve common objectives (e.g. the local administration). Lack of collaboration between the agrarian union, farmers and politicians.*
Behavior of big powers	Lack of compromise of organizations (administration, universities, great trades…) with capacity and power to promote a new urban concept that includes protection of areas which produce ecosystem services.
	Sometimes, dogmatic position of some political manager that can produce a loss of allies (e.g. housewife).
	Risk that Horta's Council will disappear when future political changes occur.
Behavior of social agents	Lack of consensus among all social agents in favor of promoting agricultural activities on the way to act.
	Agrarian union representation in Horta's Council not proportional to the importance of each union, and little weight of global representation, compared to the representation of the administration.
	Some disappointment among the conventional farming community. Expectations have been created that have not been met.
Agricultural profitability	Costs produced when farmers manage dispersed small fields.
	Weak ability to negotiate sales prices, due to smallholdings.
	Differences of interest and behavior between farmer and trader.
	Greater control of commercialization and the value chain.
	Territorial fragmentation and creation of residual fields.

Table 1: Continued.

	Weaknesses
Agrarian policy	Agrarian Development Plan too focused on organic agriculture when about 90% of production is conventional agriculture.
	Agrarian Development Plan is not being executed.
	Horta's Council budget is too oriented upon surveillance against crop theft, but little on infrastructure or other productive investments.
	Dichotomy between organic (today with little production but growing) and conventional agriculture (today a majority, but with a difficult future).
	Dichotomy between ecologic dogmatism and pragmatism about reality.
	Obsolete conventional agriculture because negative environmental impacts.
	There are no clear rules about use of chemical products in agriculture. Lack of control in it use.
	Abandoned fields are a problem (rats, rabbits, bad grass/weeds…).
	Generational farmer chain broken and aging active farmers.*
Knowledge	It is necessary to have a detailed study to know better: production, stakeholders, land grouping, field structure…
	There is an observatory, but prices, distribution and the value chain are not well known.
	Training need (crop management, sales management by internet…). There is one farmer's school, but it is inefficient, according to current needs.
	Lack of cultural synergy between the university and traditional knowledge about agriculture.
	In general, society does not understand how huerta agriculture works.*
Communication	Biased information on public TV. Need an open discussion.
Environment	Land pollution by the use of agrochemicals.
	Low quality of irrigation water.

On the other hand, Horta's Council, with membership of all stakeholders, has a public budget; and in consequence, public administration has the majority of members. So stakeholders with direct interests have not a majority. In addition, these stakeholders have different points of view. It is difficult to manage this situation. To solve this problem, it is necessary to have a proactive attitude of public administration and politicians.

The second topic is about behavior of big powers, public and private. Really, public administration includes several different organizations with different strategies; and policy is not always coordinated between them. However, if strategic actions to promote agricultural products depends on public administration, there is the risk of changes when politicians change, especially in long-term actions. On the other hand, some private businesses have the ability to change framework, especially large commercial companies that can do policies to promote agricultural local production or fix prices and conditions.

In addition, social agents not always have the same interest. In consequence, it can be difficult to have a consensus between all of them, to effect strategic actions.

Besides, agricultural production has profitability problems. This question must be qualified according to the effects of the COVID-19 pandemic. In fact, the pandemic has produced an increased demand from Spain and the EU, especially for organic products. Part of the interviews was done before the pandemic and others after confinement; so there could be different perceptions, before and after confinement.

It was to be expected that agrarian policy is an important topic. "L'Horta" produces three kinds of agricultural products: conventional, not certified organic and certified organic. Now, the more important production is conventional (around 90%); but it is the form of production with more profitability problems. Organic agriculture is today a minority, but it is growing. The pandemic is producing an increase in demand, due to changes in consumer behavior; however, the certified organic products is complicated for the smallholder. In consequence, some farmers work with uncertified organic agriculture. Moreover, the side-by-side coexistence between organic and non-organic agriculture is difficult. Maybe this is one of the main important problems to solve.

On the other hand, the possibilities of "vertical farms" to make all kinds of agricultural products, organic and not-organic, could be, in the near future, the most important challenge. This question introduces the next topic, that is, the need for knowledge about management of production with new technologies, in the global market. The Internet and applications for "smart territory" are going to change how we understand agricultural food production, especially in a smallholder situation. Traditional farmers are formed in other ways to manage both land and market. It is necessary to find a way to combine traditional knowledge with new technologies. Communications with society and about the environment close the list of weaknesses.

In Table 2, you see the synthesis of the main threats as four topics: agricultural profitability, technical innovation, new transportation infrastructures, and urban development and agrarian policy. Among them, we underline the low demand from foreign markets (before the pandemic, particularly in the case of EU markets) due in part to Brexit, a pre-pandemic general agrarian crisis and agrarian policy.

The agricultural surface on "L'Horta" is big, and the local market of metropolitan area of Valencia is not large enough to consume all production. For this reason, an important part of production normally goes out mainly to EU markets; however, due to the pandemic, the EU increased its demand for agricultural products from EU countries, for security reasons.

The demand for organic product is increasing, so that is a problem for those using conventional agriculture, which is at the time the majority of production. This problem is a threat, but also an opportunity, that requires an effort to reconvert the profile of farmers, from being a traditional farmer to becoming a professional specialized in new technologies, and control and management systems, who takes advantage of popular knowledge about the agriculture of "L'Horta".

As already mentioned, maybe the main threat is the probable evolution of agriculture to a "vertical farm" system. Despite its important energy consumption, it can be a very competitive system for agricultural production, even of organic food. In this case, it will be

Table 2: Synthesis of threats.

Threats	
Agricultural profitability	Unfair competition from third world countries with lower phytosanitary requirements.
	Agrarian crisis, with prices below the cost of production (before Coronavirus).
	Lack of a foreign market.
	Certificate of organic agriculture has rigorous controls that penalize farmers, especially when there is a mix of conventional and organic agriculture.
	The great stores can control prices and impose conditions.
Technical innovation	Probably "vertical farms" will be the immediate future of agricultural production, via hydroponic and aeroponic crops. Today it is more expensive, but the cost of production will probably go down. Can make organic agriculture with absolute control of all the parameters, in buildings or warehouses.
	Emergence of a new kind of farmer: "vertical urban farmer"
New transport infrastructure and urban development	Central Administration is planning the expansion of transportation infrastructures that already exist (motorways) or the construction of new infrastructures (high speed railway).
	New transport infrastructures produce an irreversible transformation of agricultural lands, fragmentation and may render residual fields unusable.
	Use of protected agricultural land as a reserve to implement new infrastructures or urban development.
	Urban pressure to build in protected areas.
Agrarian policy	Traditional seed chain is breaking (dependence on the foreign market for seeds).
	Lack of interest of the new generations to work in agriculture is aggravated by a lack of profitability.
	Lack of results gives risk of going backwards in the process.

necessary to find other added values, to make the agriculture of these of "L'Horta" lands competitive. Agrarian policy is important in addressing this question.

Finally, new transportation infrastructures and urban development remain a major threat. They destroy agricultural land and fragment it irreversibly. The risk of land speculation is always latent. The legal framework of land ownership in Spain facilitates speculation when the right economic conditions exist. It is possible to speculate on the land if its legal protection is changed.

Table 3 shows the synthesis of strengths. Despite all its problems, still the agricultural lands of "L'Horta" have many strengths. Having a plan and a law approved about agricultural lands of "L'Horta" is already a strength. There are many social agents to maintain, protect and promote the agricultural land of "L'Horta", with a very high valuation as a heritage site for L'Horta. There is international recognition of positive values of "L'Horta", especially about water management and irrigation. This traditionally agricultural land is a sustainable system because it has been operating without interruption for more than 1,000 years. Although not many, there are young entrepreneurs willing to promote new forms of agriculture who are more competitive and have adapted to new challenges.

Table 3: Synthesis of strengths.

Strengths	
Governance	Horta Law is key to stopping the urbanization/destruction of good agricultural soil.
Behavior of social agents	There is a strong social movement, very diverse and with a lot of initiative, in favor of Horta agricultural land. It is a bottom-up entrepreneurship system.*
	An important part of the local administration shares this movement.
	Consumer attitude change: there is a major demand for local and organic products.
	There are some groups of enterprising young farmers.
	Very positive social assessment of the agricultural heritage of L'Horta.*
	There is an effective and traditional water management community, with a Water Court. It is a collaborative and self-organizing tradition.
Environment	The Huerta is an agriculturally sustainable system that has existed there for more than 1000 years.
	The traditional irrigation system helps against climate change, the hot island phenomenon and hot waves.
	Huerta-swamp symbiosis increases raining.
	The Huerta system increases biodiversity.
Cultural	"L'Horta" is an identity referent and a cultural landscape.
Agricultural profitability	Small-scale farming has drawbacks but also advantages: diversification of crops, diversification of value chains, versatility, diversification of commercial models.

From an environmental point of view, the peri-urban agricultural land is an important green infrastructure adjacent to the metropolitan area of Valencia. It is also an element of identity to Valencian society. Small-scale farming has drawbacks, but also advantages.

Finally, Table 4 shows a synthesis of opportunities. In fact, there are many opportunities from the point of view of governance, agricultural profitability, technical innovation and behavioral changes by social agents in favor of this space.

Table 4: Synthesis of opportunities.

Opportunities	
Governance	SIPAM-FAO registration: It is an international award that can help to create a mark, promote rural and cultural tourism, create professional farmers.
	Definition and identification of professional farmers (organic and conventional agriculture).**
	Horta's Council can take initiatives and planning actions because it has the budget and the ability.
	Horta's Council is an entity open to all sensibilities.
	Promoting food sovereignty of metropolitan Valencia.*
	Make organic farming and conventional farming compatible.
	Mix inspection teams: technical experts + farmers.
	Make groups for collaborative economy in agriculture.
	Need for supra-individual organization.
	EU Recovery Plan promoted because of COVID is an opportunity to transform peri-urban agriculture.
Agricultural profitability	Promoting more competitive farmers/rural companies by increasing farm size, making origin-identified products, promoting proximity markets*, land grouping…*
	Promoting land stewardship to maintain agriculture and the landscape.*
	Carrying out a study for a long-term horizon set of years (e.g. 20 years), in order to analyze how to defend the heritage of "L'Horta" (envisioning).
	Coronavirus brought increasing demand of local products and organic agriculture in Valencia, Spain and Europe.*
	Create a direct distribution system between farmers and consumers.
	Last economic crisis (before Coronavirus) generated a limited return of workers to agriculture activity.
	Possibilities of agricultural wholesale market (MERCAVALENCIA): management of "tira de comptar" for direct sales, sales of processed products, online sales and house distribution, offering of certified organic products and traditional products (non-certified organic products).
Technical innovation	Take advantage of technological changes, e.g. by APPs to sell fresh food.
Behavior of social agents	Change of mentality in favor of local products.

4 CONCLUSIONS

Despite the great effort made to elaborate and approve the plan (2019) and law (2018) of "L'Horta" with a very high technical quality, in reality these achievements are only tools to do actions. To date, there are still no visible results. All possibilities are open. There are still weaknesses and threats, but also significant strengths and opportunities. It is urgent that the Horta Council take the initiative to develop and implement concrete actions, with visible results that generate hope among the stakeholders. Many of these actions are included in the Law of L'Horta itself. It is necessary to create collaborative work between stakeholders, to gain success against the challenges.

REFERENCES

[1]   Antolín Tomás, C. et al., *El suelo como recurso natural en la Comunidad Valenciana* (*The Soil as a Natural Resource in Valencia's Community*), Generalitat Valenciana, Conselleria d'Obres Públiques, Urbanisme i Transports: Valencia, 1998.

[2]   Boira, J.V., Urbanismo expansivo: de la utopía a la realidad (Expansive urbanism: from utopia to reality). *Proceedings of XXII Congreso de Geógrafos Españoles*, pp. 79–90, 2011.

[3]   Miralles i García, J.L., La iniciativa legislativa popular per L'Horta de València (Historical popular legislative initiatives for L'Horta of Valencia). *València, 1808–2015: la historia continua*, Balandra Edicions: Valencia, pp. 217–231, 2016.

[4]   Miralles i Garcia, J.L., Real estate crisis and sustainability in Spain. *WIT Transactions on Ecology and Environment,* vol. 150, WIT Press: Southampton and Boston, pp. 123–133, 2011.

[5]   Miralles i Garcia, J.L. & Martínez Llorens, F., Tourist development and planning on the Valencian Mediterranean coast: The case of La Devesa del Saler. *WIT Transactions on Ecology and the Environment,* vol. 217, WIT Press: Southampton and Boston, pp. 495–507, 2018.

[6]   García-Ayllón, S., La Manga case study: Consequences from short-term urban planning in a tourism mass destination on the Spanish Mediterranean coast. *Cities,* **43**, pp. 141–151, 2015.

[7]   Miralles, J.L. & García-Ayllón, S., The economic sustainability in urban planning. Case: La Manga. *WIT Transactions on Ecology and the Environment*, vol. 173, WIT Press: Southampton and Boston, pp. 279–290, 2013.

[8]   García-Ayllón, S. & Miralles, J.L., The environmental impacts of land transformation in the coastal perimeter of the Mar Menor lagoon (Spain). *International Journal of Design Nature and Ecodynamics*, 9(2), pp. 109–128, 2014.

[9]   Mateu, A. & Domínguez, M., Cuando El Saler volvió al pueblo (When El Saler was returned to the people). *Mètode*, **70**, 2011. http://metode.cat/es/Revistas/Dossiers/La-Albufera-de-Valencia/Quan-el-Saler-torna-al-poble. Accessed on: 29 May 2020.

[10] Romero, J. & Francés, M., *La Huerta de Valencia. Un paisaje cultural con futuro incierto* (*Valencian Vegetable Farming. A Cultural Landscape with an Uncertain Future*), Universitat de València: Valencia, 2012.

[11] Temes, R. & Moya, A., Dynamics of change in the peri-urban landscape of Huerta de Valencia: The case of La Punta (Valencia). *WIT Transactions on Ecology and the Environment,* vol. 192, WIT Press: Southampton and Boston, pp. 123–131, 2015.

[12] Temes, R. & Moya, A., Typology of the transformations occurred in the peri-urban space of "Huerta de Valencia": Evidence from north arch of Valencia, Spain. *International Journal of Sustainable Development and Planning*, 11(6), pp. 996–1003, 2016.

[13]   Temes-Cordovez, et al., Las huertas periurbanas del Mediterráneo (Murcia–Alicante–Valencia y Zaragoza). Primeros resultados de investigación para el caso de Valencia (Mediterranean peri-urban vegetable farms (Murcia, Alicante, Valencia and Zaragoza). First results from investigating the case of Valencia). *Proceedings of III International Congress ISUF-H*, pp. 41–55, 2019.

[14]   Miralles i García, J.L., Environmental management of peri-urban natural resources: L'Horta de Valencia case study. *WIT Transactions on Ecology and the Environment*, vol. 192, WIT Press: Southampton and Boston, pp. 99–110, 2015.

[15]   Marqués Pérez, I., Segura, B. & Maroto, C., Evaluating the functionality of agricultural systems: Social preferences for multifunctional peri-urban agriculture. The "Huerta de Valencia" as case study. *Spanish Journal of Agricultural Research*, **12**(4), pp. 889–901, 2014.

[16]   Romero, J. & Melo, C., Spanish Mediterranean huertas: Theory and reality in the planning and management of peri-urban agriculture and cultural landscapes. *WIT Transactions on Ecology and the Environment*, vol. 193, WIT Press: Southampton and Boston, pp. 585–595, 2015.

[17]   Melo, C., "L'Horta de València": Past and present dynamics in landscape change and planning. *International Journal of Sustainable Development and Planning*, **15**(1), pp. 28–44, 2020.

[18]   *Estrategia Agroalimentaria. València 2025.* Ajuntament de Valencia: València, 2018. https://hortaipoblesvalencia.org. Accessed on: 29 May 2020.

[19]   Pla d'Acció Territorial d'Ordenació i Dinamització de l'Horta de València; Conselleria d'Habitatge, Obres Públiques i Vertebració del Territori, Generalitat Valenciana. www.habitatge.gva.es/ca/web/planificacion-territorial-e-infraestructura-verde/huerta-de-valencia. Accessed on: 16 May 2020.

[20]   Ley 5/2018, 6 Mar., de la Generalitat de la Huerta de València. Generalitat Valenciana: Valencia. www.dogv.gva.es/datos/2018/03/12/pdf/2018_2459.pdf. Accessed on: 29 May 2020.

[21]   Marzal Raga, R. (ed.), *El suelo rural periurbano: Estudio del caso: L'horta de València (Rural Peri-Urban Soil: Case Study of L'Horta of Valencia)*, Editorial Aranzadi Thomson Reuters: Spain, 2019.

[22]   Miralles i Garcia, J.L., New policies for the management of peri-urban agricultural spaces: The case of "L'Horta de València" (Spain). *International Journal of Design and Nature and Ecodynamics*, **13**(4), pp. 361–372, 2018.

[23]   Pla de Desenvolupament Agrari de L'Horta de València, Generalitat Valenciana, Conselleria d'Agricultura, Desenvolupament Rural, Emergència Climàtica i Transició Ecològica. www.agroambient.gva.es/va/web/agricultura/novedades/-/asset_publisher/cDoEgHxQ2gTH/content/plan-de-desarrollo-agrario-de-la-huerta-de-valencia. Accessed on: 29 Jun. 2020.

[24]   Food and Agriculture Organization of the United Nations, Historial Irrigation System at L'Horta de València, Globally Important Agricultural Heritage Systems. UNFAO: Geneva. www.fao.org/3/ca8304en/ca8304en.pdf. Accessed on: 29 May 2020.

[25]   Tribunal de las Aguas. www.tribunaldelasaguas.org/en/. Accessed on: 29 May 2020.

[26]   Garcia-Ayllon, S. & Miralles, J.L., New strategies to improve governance in territorial management: Evolving from "smart cities" to "smart territories". *Procedia Engineering*, **118**, pp. 3–11, 2015.

[27]   Jalkanen, J., Toivonen, T. & Moilanen, A., Identification of ecological networks for land-use planning with spatial conservation prioritization. *Landscape Ecology*, **35**, pp. 353–371, 2020.

SHORT FOOD SUPPLY CHAINS IN BARCELONA'S MARKETS

XOSÉ ANTÓN ARMESTO-LÓPEZ[1], MARÍA BELÉN GÓMEZ-MARTÍN[1],
MARTÍ CORS-IGLESIAS[1] & EMILIO MARTÍNEZ-IBARRA[2]
[1]Department of Geography, University of Barcelona, Spain
[2]Department of Regional Analysis & Physical Geography, University of Granada, Spain

ABSTRACT

The relationship between the city of Barcelona and the surrounding countryside has always been partly responsible for shaping the territory. Throughout history, the supply of food to the city has been shown to be the principal vector connecting these two settings. Within the city of Barcelona, food supply was traditionally linked to its immediate hinterland. However, starting in the 1960s, with the adoption of the productivist agricultural model, the commercial range became greatly expanded. Together with other factors, this seriously threatened much of the city's agrarian hinterland. In recent years, this has started to change. Given the increased geographical mobility of the population and the lack of knowledge of the agricultural world, tendencies to revalorise the environment and the territory surrounding the city have emerged. These movements are committed to maintaining farming activity and heavily promote local consumption. In this work, we study how "local food" is represented in one of the most important examples of the traditional commercial fabric of the city of Barcelona: the municipal markets. To this end, we selected a sample of the 39 food markets in the city and identified the stalls that sell this type of produce. We established a typology of products, of traders according to their interest and involvement, and of consumers. Our results show that, although local produce is advertised in many of the stalls in the markets we visited, there is considerable variation in the concept of "local", in the quantity and type of food products on offer, and in both the commitment of the traders to local produce and the consumer demand for it.
Keywords: local food, urban municipal markets, Barcelona.

1 INTRODUCTION

For decades, many academics who specialise in the rural world have spoken of the confrontation between the globalising industrial food model and alternative food models [1]–[4]. Among the alternative proposals, commitment to more environmentally friendly and socially just agricultural production has grown in recent years. Taking these two premises into account, it has become necessary to interpret the new models of production, distribution, marketing and consumption; what we can generically call "local food". This local produce can even be considered as an example of a backlash against the prevailing industrial agri-food model [5] which is controlled by market forces. Thus, in turn, industrialised markets can be seen to control our food, which is becoming increasingly unrelated to consumers' immediate settings. As a consequence, our culinary traditions, often based on the consumption of fresh and seasonal food, are losing the importance they once had [6].

Here, we aim to show the degree to which local food products are available in a traditional commercial framework: municipal markets. Such establishments are an institution that is certainly in no way exclusive to the city of Barcelona, despite the fact that some consider it to be "the city with the greatest number of markets in the world" [7]. Indeed, there is a universal character to municipal markets: they have their own rules and are highly social places that serve different functions [8]. There are many different embodiments of this typical establishment in the city of Barcelona, which vary depending on factors ranging from the geographical location itself to the socioeconomic profile of the area they serve. In addition, we will analyse the concept of "local food" that different market traders have.

WIT Transactions on Ecology and the Environment, Vol 243, © 2020 WIT Press
www.witpress.com, ISSN 1743-3541 (on-line)
doi:10.2495/UA200021

2 METHODOLOGY

We adopt an essentially qualitative methodological approach. First, we carried out a literature search for work that addresses the subject of municipal markets and the typology of local food products, and we reviewed the results. In the light of this, we designed a semi-structured interview concerning local produce and how both market traders and consumers relate to it. Finally, the responses were collated and analysed, and we interpreted the results in order to present them.

We decided to work with just a sample of the 39 municipal food markets in the city of Barcelona. In order to achieve broad geographical coverage, we selected 10 markets: one from each of the ten city districts or boroughs (see Fig. 1 for location of selected markets). We conducted a total of 38 interviews with market traders working with different types of food (fruit and vegetables, fish, poultry, fresh meat, processed meat products and eggs). The duration of the interviews varied greatly due to the range of interest and degree of involvement of the traders. Moreover, they were conducted *in situ* on working days and so most of them were interrupted by the need to attend to customers. Despite this, most of the interviewees were very receptive and interested in the study (see Table 1 for a typical interviewee profile).

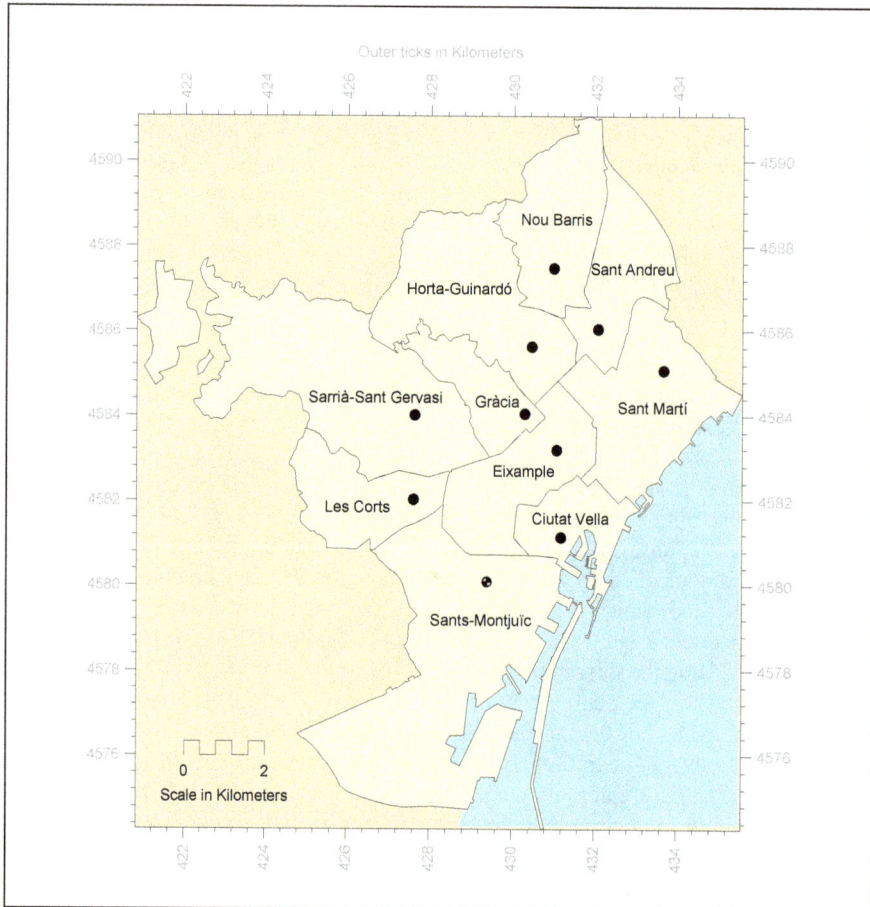

Figure 1: Barcelona's city Districts and location of 10 selected markets.

Table 1: Profile of a typical interviewee

Sex	Female
Age	46–60 years
Size of stall	11–20 m^2
Age of business	> 50 years
Number of workers	2–5
Main products	Fish and seafood, meat and meat derivatives, fruit and vegetables

3 BARCELONA, A CITY OF MARKETS

Municipal markets came into being in Barcelona over a hundred years ago in response to the growing problem of supplying demand for fresh produce within the city. Their ultimate goal was to offer the city's population a broad and reliable range of products, within the context of increased political concerns for public sanitation and hygiene. The model was very popular in the city and the municipal markets grew in importance as the population increased. This growth was the result of the city becoming consolidated as a first-order industrial and service centre. It was not until about thirty years ago that the city's markets started to fall into decline, due mainly to changes in patterns of consumption. Thus, this business model fell under threat from changes in product preferences, a timetable that restricted opening to the morning, an aging clientele linked to the massive incorporation of women into the labour market and, most importantly, competition from other operators: supermarkets, specialist shops and department stores [9].

The municipal markets in Barcelona are 180 years old, although of course food markets date from much earlier and evolved in parallel with the very first growth of the urban area. In the Mediterranean world, the agora first and the forum later functioned as places of exchange for all kinds of products. After the gradual expansion of trade over several centuries, real growth did not come to fairs and markets until the Middle Ages, when some became periodic meetings of merchants in urban centres that were well connected to their surrounding territory [10]. In the 19th century, industrialisation and urban growth inevitably led to the expansion of food markets, driven by the growing need for food. During the 20th century, markets around the world evolved very differently, depending on the city where they were and on the socioeconomic context of the country, they were in.

The first permanent market in the city of Barcelona dates from the 10th century, although the current model of municipal markets first took form in the 19th century, as laid out in the Urban Planning Ordinance [10]. This was the time when many of the city's most emblematic markets appeared: La Boqueria (1840), Barceloneta (1884), Santa Caterina (1888). From the last part of the 19th century and throughout the 20th century, numerous municipal markets came into being in the city. They would go on to have very different fates. Some became consolidated hubs of commercial or tourist activity, such as the markets of Santa Caterina or La Concepció. Meanwhile, others simply disappeared, as in the case of the markets of Vallvidrera and Carme: swallowed up by competition from renewed more flexible trading guidelines that were adapted to the timetables of workers, and the opportunities presented by improvements in local and regional transport.

Since the early 1990s, Barcelona's municipal markets have been the responsibility of the Barcelona Municipal Institute of Markets (Institut Municipal de Mercats de Barcelona). This is a self-governing body that functions as a subsidiary of Barcelona City Council. In this period, more than half of the municipal markets have been overhauled and modernised. This restructuring has involved infrastructures being updated, commercial offers being adapted, supermarkets being introduced into dedicated areas and new services being incorporated. Overall, the social and environmental functions of the markets, as well as the roles they play within the city, have been reinterpreted [10].

Barcelona's markets currently generate 7,500 jobs; they occupy a trading area of 60,901 m^2, accounting for 26.6% of the commercial offer in the food sector; and they have an annual turnover of over €950 million [7], [8], [10]. In addition, the 39 markets receive a total of 62 million visits per year. According to the latest Barcelona municipal survey, the city's markets are the first choice of consumers when it comes to buying fresh food; although they lag far behind supermarkets in more general terms of food; the preferences of buyers lie on supermarkets when food is considered as a whole category.

4 THE PRESENCE OF LOCAL PRODUCE IN BARCELONA'S MARKETS

As a result of the in-situ interviews we conducted with market traders, we can describe and interpret the presence of food from short supply chains in Barcelona's municipal markets. In this section, we consider the conception of local produce, the classification of products according to geographical range, and the strengths and weaknesses of these products in the opinion of the market traders.

4.1 Short food supply chain and local produce: traders' nuances

Establishing a single definition to refer to local produce is a rather complex task. The manifold and specific characteristics of each type of product, the range of geographical characteristics, the varied socioeconomic and environmental relationships that occur in the different territories and at the different scales in question [5], together with political ideology and the historical moment, make complete agreement on this issue almost impossible. However, the "short supply chain" concept can generate a greater degree of consensus; although, it is not without its ambiguities either.

For the topic we are dealing with here, the perspective that is introduced according to the type of establishment, and therefore the type of product, is of considerable importance. This means that the responses to the interview varied depending on whether the interviewee ran a fish stall, a butcher's stall or a fruit and vegetable stall. These were the three main categories covered in this work, as they are also the most common in Barcelona's markets. In general, we can say that for many of the respondents, the precise meaning of "local" was somewhat complicated; although we can glean some overall results. The notion of "short food supply chain" turned out to be even more complex, since the terminology was new for most of the interviewees. Notwithstanding, once the meaning was clarified, most of them considered that a short supply chain would be one that consisted of no more than three links in making produce available to consumers.

For products from the sea, on several occasions the territorial dimension of "local" was equated to speaking of the Mediterranean Sea. The most common reference was to what was the neighbouring area: the Mediterranean Sea as a unifying and defining element of this part of the world, and also of the sentiments of traders in a coastal city such as Barcelona (see Fig. 2). Many times, when asked by the interviewer "the whole of the Mediterranean Sea?" the respondent would offer a limit that was highly variable and ranged from the areas where fishing boats captured the fish sold by them (almost always, very close to Barcelona) to an idea of the Mediterranean Sea that stretched beyond Spanish territorial waters. Another frequent response was the use of the concept of "costal fish" to refer to the catches of inshore fishing. With regard to a short food supply chain, there were two classifications identified in the responses. On the one hand, there were stallholders who mostly bought their fish from Barcelona's central wholesale food market (Mercabarna) and who, therefore, introduced at least one intermediary into the supply chain. On the other hand, there were fishmongers who purchased supplies, at least in part, directly at the fish auctions organised at nearby ports by the local fishers' guilds. In the latter case, this was the main argument offered to clarify what they understood by quality; adding that arriving straight at the stall, without intermediaries, should be considered as the most important added value of such products.

Figure 2: Fish stall with the local sign. *(Source: Image by the authors.)*

When it came to fresh meat products, the responses tended not to clarify the reference scale. There was a fairly even division between those who considered "local" to refer to the whole of Spain and those who limited "local" to the region of Catalonia. The issue of the origin of meat and the extent to which it can be considered local proved to be rather difficult to resolve, and numerous variables related to the concept of quality emerged. The idea of quality was associated with geographical origin (where the highest accolades went to Galicia for beef, Aragon for lamb and Andalusia for kid), but it was also linked to the breed of the animal as well as to livestock management practices and proximity to distribution channels (feeding methods, distance to slaughterhouses, etc.). It was in this context that the concept of a short food supply chain was once again introduced; although, as we have said, most of the interviewees were not familiar with the term. However, once the concept had been clarified, they were quick to establish a threshold for considering a product to come through a short supply chain. This can be summed up in a statement made by one of the stallholders "the meat is brought to me from the slaughterhouse in Caldas de Montbui [a town 35 km from the centre of Barcelona] and that means that from the farmer, it only passes through two sets of hands before it arrives at my stall (…)".

Fruit and vegetable sellers mostly thought that "local" meant from the region of Catalonia. For this type of produce there was, however, a greater variety of local scales offered – from specifying the central coast of Catalonia to specifying the province or the metropolitan area of Barcelona. There were even traders who quantified in kilometres how far "local" stretched: one cited exactly 100 km as the limit; while another extended that distance to 200 km. Once again, conceptualisation of short supply chains was, from the start, more complicated. In fact, for some interviewees, the short supply chain was reduced to the distance between the producer and the market stall. We should note that some of the respondents owned agricultural plots and therefore channelled their produce through their own stalls and offered their personal experience as an example of a short supply chain.

4.2 Products and their origin

Almost all the interviewees stated that they sold the type of produce that we address in this work. In this section, we specify the origins given for the produce included in the three food groups applied in the previous subsection: fish and seafood; meat and meat products; and fruits and vegetables.

With respect to the establishments that sell fish and seafood, we can summarise the situation by saying that market traders defined as local produce the smaller species that some called the "seasonal catch". This term refers to the natural cycles of the appearance or of these species or of their entry into the fishing grounds near the Mediterranean coast, and includes hake, *Merluccius merluccius*; red mullet, *Mullus surmuletus*; prawns, *Aristeus antennatus*; and mantis shrimp, *Squilla mantis*. In general, they also used the names of fishing ports where the catch is landed as geographical references to limit what can be considered local: Palamós, Sant Carles de la Ràpita, Blanes, Roses, Tarragona or Benicarló. All except for the last of these are on the coast of Catalonia. As an example of how complex the identification of marine products sometimes is, with respect to the framework of territorial identity, we can cite the use of "local" when used to refer to the clam known in Barcelona as "rosellones" (*Chamelea gallina*). It is a common and popular shellfish in this area, where it is often caught in the Gulf of Lion: in French waters. However, this trader differentiated these clams from the equivalent catch in neighbouring Italian waters, in the Gulf of Genoa, which were not considered to be local. In the stallholder's words: "we could generally say that chirlas are local produce, they come from the Gulf of Lion; what happens is that there are times when the distributor at Mercabarna only has them from Italy". Therefore, in both cases, the origin was outside Catalonia and even Spain. However, as a result of the traditional shared use of neighbouring fishing areas between ports in the north of Catalonia and the south of France, there is a genuine sentiment that produce from the latter is still local.

The responses of those running butcher's stalls tended to use a variety of scales when referring to products as being of local origin. They often included regions that were some distance from the point of sale. In this way, the beef that was sold in these establishments and considered to be local ranged from a distance of just over thirty kilometres away at its closest (the town of Granollers and the Alt Penedès area were explicitly mentioned) all the way to the region of Navarre. So, as we have already said, conceptual differences were more than evident (see Fig. 3).

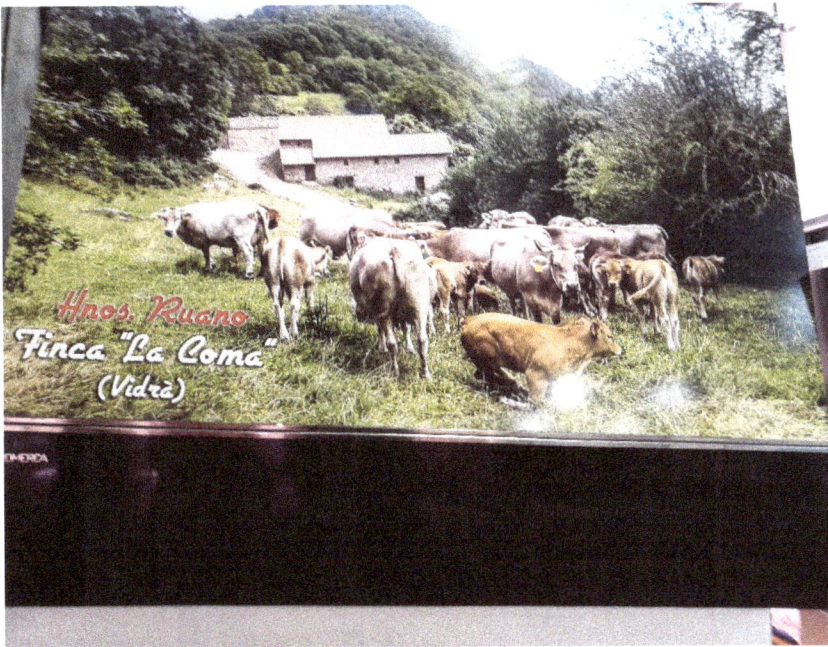

Figure 3:   Sign in a market stall identifying the origin of livestock in the village of Vidrà, 105 km from Barcelona. *(Source: Image by the authors.)*

In the case of lamb, the situation was somewhat different. Here, no specific Catalan towns or areas were included in the responses. When stallholders mentioned local produce, they generally referred to places outside Catalonia, with relatively frequent remarks concerning Aragon: a region that borders on Catalonia. Only occasionally did generic responses mention Catalonia. There may be several factors that explain this difference. One would be the lack of a tradition of eating the local sheep breeds in Barcelona. Another is the preference for other Spanish breeds of lamb that are popular in their places of origin and are requested by clients originally from those regions who moved to Barcelona many years ago. Finally, there has been a tendency for sheep and goat farming to be substituted by beef farming throughout all the northern mountain ranges of Spain over the last century or so.

When considering goat meat, the origin was almost never classified as local: only one informant considered Almería (about 550 km from Barcelona) as local. The reasons given for this are similar to those in the case of lamb but are even more pronounced. There is only one breed of goat native to Catalonia, the Rasquera White, which is found in mountainous areas in the south of Catalonia and is currently endangered [11]. Moreover, in the city of Barcelona the consumption of goat meat has always been limited. Finally, regarding fresh pork, we can only say that the interviewees offered few responses referring to local produce, with the province of Girona being the most common response.

With regard to greengrocers and fruit stalls, we note that this is the commercial sector in which the respondents suggested the narrowest range for "local". Most focused on the proximity of the products in the regions surrounding the city of Barcelona, frequently mentioning Maresme (see Fig. 4), Baix Llobregat and Vallès Occidental. The names of towns within the Barcelona metropolitan area where some horticultural and fruit produce has a very good reputation among consumers were also mentioned frequently: artichokes from El Prat de Llobregat or cherries from Sant Climent de Llobregat, for examples. At a farer scale, other place names were also mentioned, especially the neighbouring provinces of Lleida and Girona; these references were almost always limited to fruit (particularly apples and pears). For citrus fruit, the limit increased to the south of the province of Tarragona and the north of the province of Castellón, immediately south of Catalonia. We should say that normally the origin of this type of produce is not displayed at the market stalls, and it is only if the customer asks the stallholder that they can know where it is from.

Figure 4:   Tomatoes in a market stall with identified origin in Mataró municipality, 35 km from Barcelona. *(Source: Image of the authors.)*

As we mentioned above, some of the interviewees told us that they had they own farms or plots in neighbouring areas, and they commercialised their produce through their own stalls. In this case, the traditional horticultural zones in the Barcelonan conurbation delimited perfectly the area of production: the delta of the River Llobregat, the coastal plains and foothills of the coastal mountain range in the Maresme, and the plains of the Vallès basin.

4.3 The pros and cons of local produce and of short supply chains for traders

In the last part of the interview, we asked stallholders for their opinion of local produce. For this, the interviewees were required to list local produce and to comment on the advantages and disadvantages they thought it had, if they so wished. Broadly speaking, and without distinguishing between types of produce, the advantage that was without a doubt most commonly mentioned was that local produce is fresher. In contrast, the most common negative concept was its elevated price. We should also note in terms of a general overview that almost a third of the interviewees did not mention any inconvenience at all when talking about local produce.

By category of establishments, fishmongers placed emphasis on a concept that is always closely related to the world of fishing and fish: freshness. This can therefore be seen as an indispensable quality when it comes to assessing the quality of fish and seafood. At the opposite end of the scale we had the higher prices paid for products from local Catalan fishers; some respondents said that it was twice the price of the completion.

For the butchers, the most commonly cited advantages of local produce referred to the quality of the meat. For many of them, local meat is of higher quality and that has an indirect impact on precisely what was considered the most common negative aspect: the higher price of this type of produce. They positively assessed the fact that closer production reduces the time that the different meat products need to be in refrigerated facilities. Thus, making the supply chain shorter had a positive effect on the quality of the produce and this resulted in consumer safety and satisfaction.

Among fruit and vegetable traders, there was a widespread idea that the freshness of local produce is the most important positive factor that singles it out. One of the interviewees referred to this in the following terms: "… the quality of the fruit we sell here [referring to their own stall] cannot be compared with that of fruit sold in supermarkets, which undergoes many more processes of manipulation after leaving the orchard or field; the fruit we have at the stall is fresher, we treat it better". Many of the stallholders also mentioned the element of trust that local produce generates in their customers, who generally associate local fruit and vegetables with increased quality. On the downside, these traders talked of insufficient quantity of some products to cover the growing demand for them, and also the seasonal variability of some.

5 CONCLUSIONS

In broad terms, from this work we can conclude that, after going through difficulties at the end of the last century, Barcelona's municipal markets seem to have found their place within the city's complex commercial fabric. To reach this point, over recent years different City Council institutions have designed a series of strategies for the markets. These include differentiation of the products on offer, through promoting both proximity to the consumer of businesses and traders, and local food, which has experienced different degrees of acceptance and commercial success among both traders and consumers.

Analysis of this last issue shows how the concepts used by traders to refer to what constitutes local produce are varied and involve a degree of complexity that on occasion is rooted in aspects as personal as one's individual political beliefs (which were mentioned by some of the respondents) or one's philosophical outlook on life. The allusions to models of sustainability or the use of inclusive and exclusive distance scales, as well as specific place names, offer good proof of this. The concept of short food supply chain is not used much; in fact, all the interviewees referred to the issue as concerning proximity or local produce. However, delimitation of a short food supply chain turns out to be less problematic than setting a numerical or administrative distance limit on what counts as "local".

The type of product we are dealing with influences to a great extent, not only the geographical range, but also the very conception that traders have of what counts as local produce. Within the commercial spheres of fishing and of fruit and vegetables, many different references are made to what is local. In contrast, meat stallholders seem to adopt reference scales that corresponded more closely with formal provinces or regions that can be easily identified by most of the population. It could also be inferred from the interviews that a short supply chain is more easily differentiated in the case of meat and fruit and vegetable products.

We can conclude that for most stallholders, local produce is synonymous with fresh products. For the typical trader, the most commonly accepted meaning of "quality" involves precisely products being as fresh as possible and also being accepted by the customers. Thus, local produce would be high quality fresh products. Despite this general consideration, some different opinions were also expressed to justify the use of products from places that may be much further away, and always with very high standards. In these cases, however, the products were popular and sought out by clients, especially in meat establishments when referring to certain types of meat, and also in the case of fishmongers, very clearly referring to shellfish.

ACKNOWLEDGEMENTS

This research took place within the framework of the Spanish Programme for Research, Development and Innovation (Programa Estatal de Investigación, Desarrollo e Innovación) addressing social challenges, as laid out by the Spanish Ministry of Economy, Industry and Competitiveness (Ministerio de Economía, Industria y Competitividad), with reference number CSO2017-88935-R; and the Grup de Recerca Consolidat 2017-SGR-25 of the Generalitat de Catalunya.

REFERENCES
[1]   Marsden, T., Exploring a rural sociology for the fordist transition. *Sociología Ruralis*, **32**(2/3), pp. 209–230, 1992.
[2]   Bowler, I., Agricultural land use and the post-productivist transition. *La Investigación Hispano-Británica Reciente en Geografía Rural: Del Campo Tradicional a la Transición Post-Productivista*, eds A. López and F. Molinero, AGE: Murcia, pp. 179–187, 1996.
[3]   Hervieu, B., *Los Campos del Futuro*, Serie Estudios, n° 118, Ministerio de Agricultura, Pesca y Alimentación: Madrid, 1997.
[4]   Ilbery, B.W. & Bowler, I., From agricultural productivism to post-productivism. *The Geography of Rural Change*, ed. B. Ilbery, Longman: London, pp. 57–84, 1998.
[5]   Armesto López, X.A. & Gómez Martín, M.B., Restauración local y productos alimentarios. La situación en la comarca del Moianès (Cataluña). *Ager. Revista de Estudios sobre Despoblación y Desarrollo Rural*, **21**, pp. 43–72, 2016.
[6]   Such, X. et al., Alimentar Barcelona. Una Mirada crítica al model alimentari de la ciutat. ARAGUAB, Universitat Autònoma de Barcelona. http://universidadesdelatierra.org/ca/proyecto/alimentar-barcelona/. Accessed on: 18 Apr. 2020.
[7]   Mercats, Ajuntament de Barcelona, L'experiència de Barcelona; Institut Municipal de Mercats de Barcelona. https://ajuntament.barcelona.cat/mercats/en/content/ markets-barcelona-experience-1. Accessed on: 21 Jan. 2020.
[8]   Ajuntament de Barcelona, Hàbits de consum i polaritat comercial comercial de la ciutat de Barcelona. http://ajuntament.barcelona.cat/premsa/wp-content/uploads/ 2018/04/Informe-H-i-P.pdf. Accessed on: 14 Jun. 2019.
[9]   Guinjoan, M., Taula rodona: els nostres mercats són sostenibles econòmicament? Són competitius? *Diputació de Barcelona, 6è congrés de mercats Municipals de la Província de Barcelona, El Compromís dels mercats del futur*, Diputació de Barcelona y Ajuntament de Barcelona, 2014.
[10] Martin, O., (Coord.), Els mercats de la Mediterrània. *Models de Gestió i Bones Practiques*, Institut Municipal de Mercats de Barcelona, 2013.
[11] Fernández Rodríguez, M., Gómez Fernández, M. & Delgado Bermejo, J.V., *Guía de Campo de las Razas Autóctonas Españolas*, eds, S. Adán Belmonte & M. Jiménez Cabras, (Coord.), Ministerio de Medio Ambiente y Medio Rural y Marino: Madrid, 2009.

GEOMETRIC FRAGMENTATION OF THE HUERTA OF VALENCIA, SPAIN: TOWARDS A SUSTAINABILITY MODEL

ALFRED ESTELLER AGUSTÍ[1], ISAMAR ANICIA HERRERA PIÑUELAS[1]
& ADOLFO VIGIL DE INSAUSTI[2]
[1]Atemajac Valley University, Mexico
[2]Polytechnic University of Valencia, Spain

ABSTRACT

The history of Valencia's Huerta Sur through a period of 91 years (1928–2019) and the specific area of Castellar, Valencia, valued for its low physical alteration over time, composes this analysis. The focus is on understanding the composition and, therefore, fractioning of the structure of Valencia's Huerta Sur. This study presents an understanding of the geometry of the plots, from historicism related to hereditary processes, and reflects on sustainable management and economic viability of agricultural heritage. The historical evolution of the transitions and the legislation governing the land introduces criticism of the operation of Valencia's Huerta and its fragmented functionality. Using illustrative maps, the proportions and portions produced by centuries of division that modified the structure of Valencia's Huerta configuration are explored. The presentation of new forms of organization of the Huerta is a necessary reaction as a response to the conditions of abandonment and disuse of the land, which in turn conveys the current impossibility of self-management.
Keywords: sustainable food systems, geometry, fragmentation, Huerta.

1 INTRODUCTION

Taking as a starting point the justice and fairness of the succession of property, the possibilities for the transfer of heritage from a historical analysis have evolved in many perspectives, and it is difficult in some cases to define forcefully whether they have been entirely right. Specifically, Valencia's Huerta has been physically and functionally impacted during the last 100 years, as a consequence of years of hereditary processes that partitioned and continues partitioning the land to unsustainable extremes. The relationship between the balance of the natural ecosystem and the economic operation exceeds the point of equilibrium, leading to a major problem related to the disinterest and abandonment of the land, whose problem returns to the initial situation related to the way in which humanity grows. The Castellar area is one of the few areas of Valencia's Huerta near the capital that remains highly unchanged, and therefore becomes an area of interest sampling for comparative criticism.

2 THE HUERTA IN THE VICINITY OF CASTELLAR-OLIVERAL

"To the hegemony of collective or semi collective forms of ownership – the rights of use – follows the emergence of feudal forms of property, which undoubtedly involve a step forward on the way to privatization. In this manner, as cultivated areas expand, property rights tend to prevail over those of use and, at the same time, land begins to build a more important part in noble fortunes" [1]. The land has been transformed into small portions owned by many, which speaks of a democratized agricultural space described by Giobellina [2] as an essential characteristic of the Huerta. A central theme that defines the Huerta is what has happened and what is still happening with land ownership and the size of plots. Historically, the Huerta has undergone stages of concentration and de-concentration of land ownership, but there have never been large landowners. Between the 13th century and the end of the 20th century,

WIT Transactions on Ecology and the Environment, Vol 243, © 2020 WIT Press
www.witpress.com, ISSN 1743-3541 (on-line)
doi:10.2495/UA200031

property was continually subdivided, until eventually reaching the size of smallholding in which the average surface area of the plots in 1989 was 0.39 ha.

It is clear that the size of the plots has been gradually reduced in the fields, but in this case the subdivision process registers patterns that are repeated and that will tend to be repeated again and again, putting the future of this productive area in check and becoming an amalgam of plots linked with self-production. In the case of the village of Castellar-Oliveral, the portion of land located in the study area has a particular casuistry; it is one of the few unaltered areas with an exemplary cartography, a priceless map, which shows the cadastral property at the end of the 1920s, and that encompasses the area under study. This document, as well as the unaltered situation of this territory, which has been exceptionally maintained until now, makes it possible to proceed to a detailed study of the fragmentation rhythms of the Huerta plots.

3 GRAPHICAL ANALYSIS OF THE EVOLUTION OF THE FRAGMENTATION OF THE TERRITORY OF VALENCIA'S HUERTA IN THE VICINITY OF CASTELLAR

In the article "La imposible igualdad. Familia y estrategias hereditarias en la Huerta de Valencia a mediados del siglo XVIII" (The impossible equality. Family and hereditary strategies in the garden of Valencia in the mid-18th century) [3], an example is given in order to understand the distribution of land, the number of partitions related to the fifths and thirds in the inheritance of land. "We can get an idea of the territorial fragmentation (…) if we observe that the 29 *hanegada* [~24 ha] lots that the father had, split over 6 different areas, had to be divided into 10 smaller lots to cope with inheritance partition. In this case, however, it did not present too many problems since there were only two males and there was enough land to divide".

Performing a graphical analysis aims to be able to obtain close deductions by understanding plots as a metric parameter of the evolutionary conditions of land ownership and surfaces.

3.1 Determining the mapping tools

Information within the public domain, in the case of Valencia, is a tool that magnifies the possibilities of analysis. The information, freely accessed on the part of cadaster of the administrative record of the State, is capable of providing the oldest plan of the city, dated 1928, as well as the most recent dated 2019, giving a chronological window of 91 years. Although there are other graphic representations available, for example one provided by an American flight in 1948, they do not have the quality to carry out the analysis.

3.2 Maps and portions

With the purpose of leading the research by graphic interpretation, a graphic representation system is modeled that analyzes, in a chronological way, the quantity and size of the resulting portions over time so that, finally, a story can be narrated as a result of the mathematical assumption of the surface area. A section of the eastern area of Castellar is the base sample, delimited by the urban area to the west, further delimited by the V-31 highway, and Alfafar administrative territory to the south. While to the east, the human made limit is the CV-500 road, running parallel to the coastline, and the new riverbed of the Turia, built in second half of the 20th century, provides the northern limit (Fig. 1).

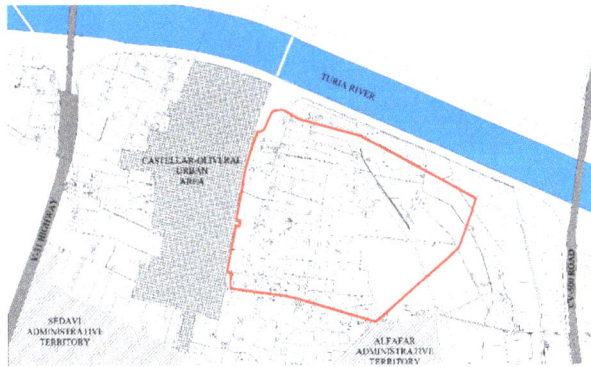

Figure 1: Castellar area site to be analyzed.

Figure 2: Plot divisions in the period 1928–1999.

3.2.1 Contrast between 1928 and 1999

Bubble map generation considering the lines that appear as a result of the division of the plots over a period of 71 years, each new line is represented by a circle whose diameter corresponds to the size of the dividing line.

The area shaded orange in Fig. 2 was converted into an industrial park during the second half of the 20th century. This had a great impact both on the functioning of the area with respect to the introduction of a new use, and because of the change in the type of ground surface and the impact on the thousand-year-old irrigation network.

3.2.2 Comparison between 1999 and 2019

Bubble map generation considering the lines that appear as a result of the division of the plots in a period of 20 years, each line is represented by a circle whose diameter is the size of the dividing line (Fig. 3).

Proportionally, in comparison with the previous map, the number of divisions that appear is substantially fewer. Due to the division of limited surfaces, as well as the partial colonization of building, since many of these divisions, particularly those that correspond to the smallest parcels, are due to the division with the purpose of building small houses on their own land, dividing the parcel.

Figure 3: Plot divisions ranging from 1999–2019.

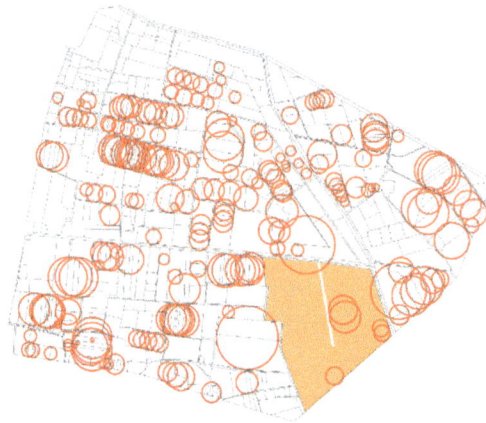

Figure 4: Cumulative divisions of plots ranging from 1928–2019.

3.2.3 Comparison between 1928 and 2019

Complete map with the accumulation of partitions in the plots over the course of almost a century – 91 years (Fig. 4).

It can be seen by combining both maps, that the fragmentation of the plots has been a generalized and constant fact over time, since it is even evident that in those areas where there were no changes during the last century, the plot sizes are already relatively small, which indicates that the land was probably divided prior to 1928. Although this fact does not exclude them from relevant observation and reflection.

3.2.4 Analysis of surfaces with regular distribution

The sizes of the plots are mapped in four ranges, with plots of less than an 831 m^2 (dark brown), plots of less than 3,324 m^2 (orange), plots between 3,324 and 6,648 m^2 (yellow), and then larger plots shown green, the larger areas being darker (Fig. 5).

The aim is to obtain a generalized distribution, in which the larger plots stand out.

Figure 5: Zoning by surface size in arithmetic ratio.

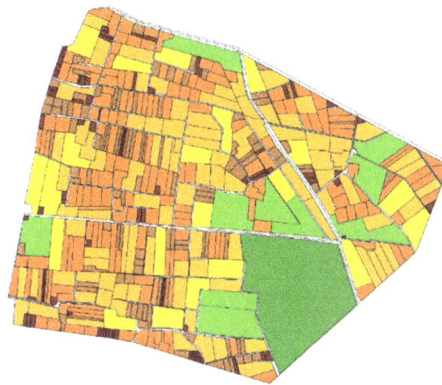

Figure 6: Zoning by surface size in geometric relationship.

3.2.5 Analysis of geometrically shaped surfaces

The previous surface area analysis highlighted the need to be able to distinguish between those plots of smaller size, since they are the ones that best illustrate the processes of fragmentation. With this objective, a new map was generated in which the distribution of surfaces follows a geometric relationship, doubling the previous surface, based on the hypothesis that fragmentation is carried out at least between two parts, together with the analysis of adjacent surfaces in which the relationship between them was mostly 1:1 or 1:2 (Fig. 6).

There is obviously some coincidence with Fig. 3, in that those areas that have been divided in the last 91 years appear with darker brown colors. However, it is also easily observed that the division into smaller plots is not only located in those areas that have been subdivided in recent years.

3.2.6 Analysis of divisions

In order to be able to determine a historically altered region, the plots that were divided, in orange, during the period 1928–2019 from those that remained unchanged during that period, in green, are represented in a contrasting way (Fig. 7).

Figure 7: Identification of divided and not divided plots.

This classification shows that this fragmentation repeatedly began at least more than 200 years ago, especially when observing the level of fragmentation of some green areas, and together with the calculation of the time necessary to reach that level of subdivision. In the first decades of the 19th century [4], the extension of the Huerta amounted to approximately just over 357.33 ha distributed in about 1,600 owners. We could think of an average surface area of 2.7 hectares per owner, however, the distribution was very uneven. Of the total of the owners, 36% had less than 415 m^2, which represented 2.9% of the total area. If we move to the opposite extreme, 6.2% of the owners had 51.3% of arable land, which accounts for the accumulation of land by large landowners. The nobility (61 owners), the clergy (62 owners) and the bourgeoisie (549 owners) accumulated 86.1% of Valencia's Huerta with average properties that ranged between approximately 1.16 ha for the first two, and 30.74 ha in the case of the bourgeoisie. The rest, owned by the peasantry, corresponded to a total of 74.57 ha. It should be noted that the income per hectare was very similar among all social groups and ranged between 348 and 396 pounds/ha. Large plots were numerous in the case of large landowners, with plots between 1,000 and 4,000 m^2, in more than 50% of the cases, while in the case of the peasantry, 50% of the cases remained between 250 and 1,250 m^2, with plots of less than 250 m^2 being 20%.

3.2.7 Specific analysis of the division of plots linked to generational history
The plots with geometric division behavior are sampled to be able to translate it into historical interpretation of the partitions.

The results of a study carried out in one of the central areas of the general study area, in which the divisions suffered over time, have been traced for an area of almost 70,000 m^2. In this area, two parallel analyses of the evolution of fragmentation has been carried out, starting from the current state, towards the past and from the old cartography presented towards the present, identifying the points of coincidence and establishing the corresponding patterns both in the direction of the division and in the surface relationship.

The area was characterised by several aspects that are interesting for the present study. It is delimited in its long sides, north and south, by roads that include ditches, which makes the supply of water possible for irrigation by these two sides, facilitating the flexibility of subdivision. To the east, it borders the industrial estate, separated by an old road in existence prior to the industrial qualification, presenting an exact limit from which it also has its previous subdivision, which is important when presenting and understanding the previous irrigation lines and divisions. To the west, plots border the current urban area, which, despite not showing significant growth in recent years, are modified spaces and do not respond to the general surface plot of the Huerta.

The inclusion of different non-agricultural uses in this area also positions it as one of the most endangered areas for traditional vegetable plots. These uses are, in addition to the adjacent industrial park, a service and workshop area in one of the plots with a building next to the road. As well as the use of the plots by some owners as a warehouse for trailers and transport vehicles, waste collection or even apparent abandonment of them.

Regarding the fragmentation, the objective of the study, the one found in this area also presents varieties and combinations representative of the area in general, as well as a variety of composition that ranges from very simple to more complex divisions.

It is important to define the limitation of this study, which depends directly on the available cadastral information and the definition of the years in which the divisions were created. As previously observed in Figs 5 and 6, although the rules of subdivision may be the same in different areas, this does not mean that the time in which the subdivision is made is the same. To represent evolution in this study, with this limitation and with the objective of simplifying its understanding, fragmentation scales are not presented in temporal order between the different areas. It is possible that some plots were quickly divided with respect to others and then, they remained stagnant for a while, as the others were divided, progress of division that is subordinated to several processes and factors, economic, demographic and of inheritance, among others, that are outside the scope of the present study.

Fig. 8 shows the first large division into three equal parts, although greater divisions in the surrounding areas cannot be ruled out, while Fig. 9 shows three different proportions of division. The green plot is divided into two areas of equal surface plus the area that includes the building next to the road, this area in the future will be divided and will generate the access road to the rear plot (Fig. 10).

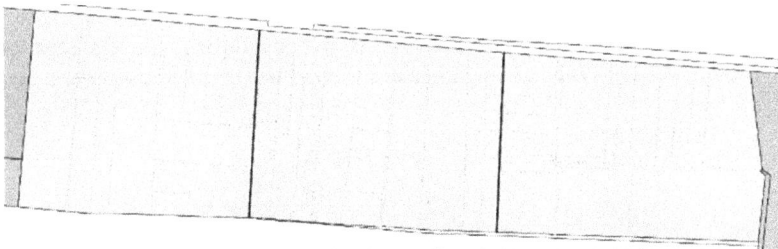

Figure 8: Zone displays with first-order partitions.

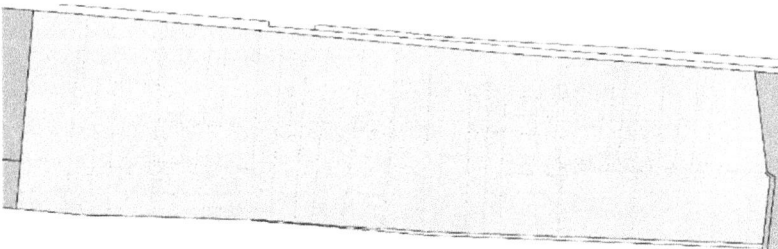

Figure 9: Analyzed zone without partitions.

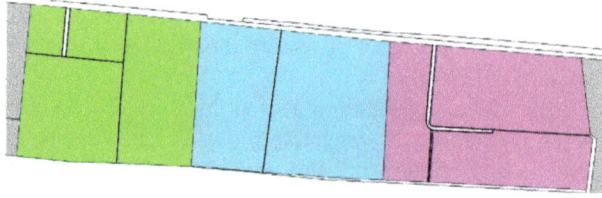

Figure 10: Zone displays with second-order partitions.

In the blue area, a second type of division appears, in fifths, "the Castilian testator divided the inheritance in five parts, four of which must necessarily be transmitted to their descendants, of them two parts must distribute them equally among children and grandchildren and the remaining part among whoever he considers appropriate: it is the so-called improvement third. The fifth part that is missing to complete the inheritance, is what we know as the free disposal fifth" [5], leaving the western part two fifths and the right part three fifths. But it is in the pink plot where the division of greater complexity is found, since between the division located in the west and the southeast appears the relation of two and three fifths respectively, while between the southeast and northeast plot the same relation of two and three fifths respectively appears again. However, the process that led to this subdivision is not known, and it also includes another interesting factor of division such as the inclusion of the east–west line to the early process of division, which corresponds to the subdivision of other plots that disappeared with the creation of the industrial park.

Fig. 11 shows new divisions in practically the entire area, except for the northwest where the first order division is intact. The most repeated divisions are in two equal parts, although there is one in three and one division in a two-to-one proportion. It is also worth noting the appearance of an access road in the horizontal division of the eastern part of the green plot.

In Fig. 12 you can see the areas that remain without further division, starting to generate the difference in areas. In this step two divisions of two and three fifths stand out, one in the second division of the blue zone and another in the southwest corner of the pink zone.

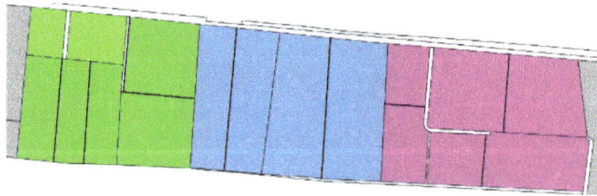

Figure 11: Zone displays with third-order partitions.

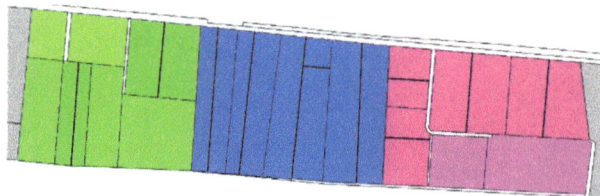

Figure 12: Area shows with fourth-order partitions.

In the divisions in Fig. 13, where a greater variety of surfaces is generated, the union to the west of the initial green plot stands out, while to the northeast of the same plot, one plot is divided into two equal parts and the other into two and three thirds, both cases with the sub-parceling of built up spaces, the squares of smaller size. In addition, apart from most divisions into two equal parts, there is a division into the same step of four parts to the east of the first order blue division, in which the proportions are 3 and 3/10 between the largest and smallest plots. In the pink area we find two subdivisions in equal parts and one in thirds.

Fig. 14 shows three new divisions in two equal parts and one third to the south center of the area corresponding to the building subplot. In this final state, we see the reduction of divisions, as well as the current variety of surfaces, in which the larger plots were generated in the third order division.

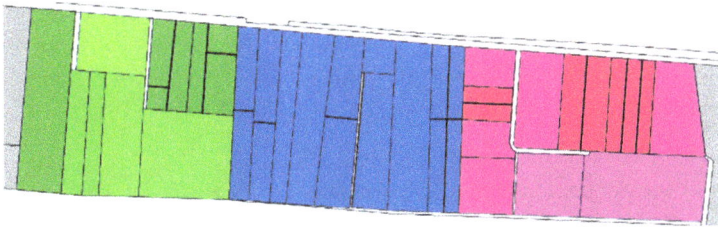

Figure 13: Zone displays with fifth-order partitions.

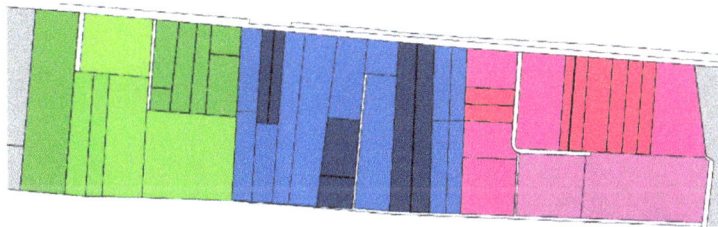

Figure 14: Zone displays with sixth-order partitions. Sexto.

4 EVALUATION OF THE CURRENT AND FUTURE SITUATION OF THE HUERTA

4.1 Critical data analysis

The fragmentation of the agricultural territory together with the plurality of owners has produced the growth in the number of plots. Having this situation effects the cost ratio maintenance – productivity – surface, becoming an uneconomic equation. This is in addition to the other problems already known around the Huerta such as aging owners and the lack of generational relief. The president of AVA-ASAJA, Cristóbal Aguado, explains that "the dramatic profitability crisis that is effecting most of the Valencian crops is created by the unprecedented aging of the agricultural population, the lack of generational support and the increasing abandonment of the fields. The alarms do not stop ringing, the numbers could not be more evident, but the government of the Botànic continues without reacting, it ignores the farmers and prefers to be in other things" [6].

4.2 Public effects of the Horta situation

Reviewing the agricultural statistics of the National Statistical Plan for the period 2009–2013, a critical aging of the population is observed, as well as a change in the trend in consumer policy that tends towards the trade of large land areas. The increase in the sale of vegetables in these establishments in the period 2003–2013 exceeded 8%, it must be remembered that these large areas are not exclusively fed by Valencian production, which unequivocally indicates a worsening of the economic situation of the farmers.

However, a review of the ages of the farmers of the Valencian garden shows that out of a total of 5,976 existing farmers in 2009, 2,456 of them, 41%, were over 65 years old, an issue that suffered a worsening in the last decade. Since in 2009, 68% of them were over 55 years old. This clearly indicates the lack of profitability that the field plots represent, given the great physical requirement that their cultivation needs that can be increasingly hard for an aging population to meet. In the same way, they emphasize an imminent subdivision with its corresponding fragmentation in the coming years as the new inheritances are incorporated into the territory.

5 RESULTS

Complementing the graphical analysis, arises the need to understand the proportions between the division of the plots and their surface area, so a statistical study was carried out with the aim of relating the area of each plot with its division process. The summary result of this study is shown in Fig. 15, in which the initial state of the graphical analysis was obviated in order to give more detail to subsequent evolution.

(a)   (b)

Figure 15: Graphs of analysis.

Fig. 15(a) highlights the descending geometric evolution for the average area of the plots, expected and corresponding to the splitting process, on the one hand, the most interesting is the change of curvature of the line of number of plots in the smaller areas, since it indicates a reduction of divisions.

On the other hand, in Fig. 15(b), the heterogeneity of the plots in terms of their surface area can be noted. This is illustrated by looking at the maximum and minimum area of the plots in each order and as the reduction of the area of the smaller plots is generated especially at the beginning of the division, generating the greatest difference between maximum and minimum plot sizes in the second order. Larger plots follow an evolution that corresponds practically to the geometric reduction of average surface area seen in Fig. 15(a), reaching horizontality in the last three orders because there are plots that were no longer divided.

6 CONCLUSIONS

The Huerta is the closest thing to a sustainable territorial model since it has survived over the centuries. The management of water resources, the preservation of the soil and the care of the plant cover, as well as the implementation of the dispersed habitat, have addressed the limiting environmental factors. A historical landscape produced by man's perseverance in the face of nature [7].

A landscape in constant transformation, a landscape that, like all landscapes, is transformed into cycles as an expression of functionality and ecosystem health, and although its inherent biological value prevails, it is also necessary to reconcile it with its historical contribution to human development. The transition from the Huerta to an economically viable model must be aligned with a series of considerations that allow for the construction of a pragmatic legacy, a heritage of cultivation.

> "The small farm can be seen as an advantage for ecological agriculture, for diversity; the Valencian case is very particular" (NGO).

> "For an integrated biodynamic farm a minimum of 4.000 m^2 is needed. (1,000–1,500 m^2 of vegetables, more fruit trees, more small animals), sold locally. A three-dimensional vegetable plot" (farmer) [2].

Four fundamental actions are highlighted that can contribute to the survival of the Huerta:

1. Regulations restricting the subdivision of the Huerta by means of protective lines.
2. It will have to establish limits by area mainly by means of biological parameters and also taking into consideration the efficiency in the relation of plot size and required installations.
3. Group land management. Whether in a collaborative or cooperative manner, maintaining decision making on land management in a collective manner can be beneficial in making the use and production of the vegetable plots more efficient. In addition, committing as part of a group rather than as an individual can bring about greater shared responsibility, committing all owners or, where appropriate, tenants to be active participants in the land and not to let it fall into neglect.
4. Expansion of the Huerta. Although it might seem a drastic measure, we must not forget that the city has stolen land from the Huerta year after year, so the re-expansion of the Huerta should be no more than a necessity corresponding to the increase of biological resources in the face of population growth, not only talking about the exponential growth of the owner families but also about what the Huerta means as a food source for humanity.
5. Modification of marketing strategies. Systems that promote the reduction of costs in the delivery of the product, associated with links with marketing companies that also allow the origin of the product to be indicated "From the Huerta" could supplement the extraordinary costs that the maintenance of the Huerta in a low surface area entails. The quality seals are one of the options in other countries of the European Union, because it allows to justify a low production in relation to the high quality of the product.

REFERENCES

[1] Perez, M., Mayorazgo y desvinculación en el municipio de Murcia, 1750–1850. https://revistascientificas.us.es/index.php/HID/article/viewFile/5915/5246. Accessed on: 21 Aug. 2020.

[2] Giobellina, B., *Procesos emergentes: de la Huerta Andalusí a la Huerta agroecológica del siglo XXI. La Huerta de Valencia*, 2nd ed. Un Paisaje Cultural con futuro incierto. Publicacions de la Universidad de València: Valencia, 88 pp., 2012.

[3] Garrido, E., La imposible igualdad. Familia y estrategias hereditarias en la Huerta de Valencia a mediados del siglo XVIII. *Boletín de la Asociación de Demografía Histórica*, **X**, pp. 83–104, 1992. https://dialnet.unirioja.es/servlet/articulo?codigo=103990. Accessed on: 20 Aug. 2020.

[4] Romero, J., *Propiedad agraria y sociedad rural en la España mediterránea. Los casos valenciano y castellano en los siglos XIX y XX*. Serie Estudios. Servicio de publicaciones agrarias. Ministerio de Agricultura: Madrid, 1983.

[5] Lagartos, F., La "mejora" como una forma de corregir el igualitarismo castellano. Comarca de Sahagún, siglo XVIII. Universidad de León. file:///C:/Users/AION%20PLAN/Desktop/Dialnet-LaMejoraComoUnaFormaDe CorregirElIgualitarismoCaste-1392738.pdf. Accessed on: 20 Aug. 2020.

[6] Furió, J., La huerta valenciana lidera el abandono de tierras de cultivo en toda España. A horta Noticias. www.hortanoticias.com/la-huerta-valenciana-lidera-el-abandono-de-tierras-de-cultivo-en-toda-espana/. Accessed on: 20 Aug. 2020.

[7] Diez, I., La huerta de Valencia. Estructura y paisaje. Territorio. https://upcommons.upc.edu/bitstream/handle/2099/13238/Palimpsesto%2004%206%2 0Igancio%20Di%CC%81ez%20Torrijos.pdf. Accessed on: 18 Aug. 2020.

SECTION 2
REDUCTION OF
URBAN HEAT ISLAND

URBAN HEAT ISLAND MITIGATION DUE TO ENHANCED EVAPOTRANSPIRATION IN AN URBAN GARDEN IN SAINT PAUL, MINNESOTA, USA

GASTON SMALL[1*], IVAN JIMENEZ[1], MICHAEL SALZL[1] & PALIZA SHRESTHA[2]
[1]Department of Biology, University of Saint Thomas, USA
[2]Department of Ecology, Evolution, and Behavior, University of Minnesota, USA

ABSTRACT

As a result of extensive urban development coupled with warming temperatures, urban heat islands (UHI) have become an important factor affecting energy consumption and human health in cities. Prior research has shown that evapotranspiration (ET) from urban vegetation can have a significant cooling effect, but there are relatively few direct measurements from urban vegetable gardens. We compared hourly temperature measurements during two summers (2017 and 2018) in a 750 m^2 research garden at the University of St. Thomas (Saint Paul, Minnesota, USA) to hourly temperatures at the nearby Minneapolis-Saint Paul (MSP) International Airport, located 6 km to the south. We also quantified seasonal ET (June–October) in 132 garden plots and five reference turfgrass plots during the summers of 2017 and 2018. For both years, an increase in temperature of 1.00°C at the MSP airport resulted in an average increase of 0.55°C in the research garden. At temperatures greater than 22°C, the garden was cooler on average compared to MSP airport. ET in the garden plots was significantly higher than in the grass reference plots both years, with means of 46 cm for garden plots compared to 19 cm for grass plots in 2017, and 51 cm for garden plots compared to 33 cm for grass plots in 2018. These results are consistent with other research showing potentially large benefits of cooling through ET from urban gardens that are primarily aimed at crop production.
Keywords: urban agriculture, vegetation, temperature.

1 INTRODUCTION

As a result of extensive urban development coupled with warming temperatures, urban heat islands (UHI) have become an important factor affecting energy consumption and quality of life in cities. Reduced vegetation cover in urban areas relative to the surrounding region results in decreased evapotranspiration (ET), causing a reduction in latent heat consumption and thereby leaving more energy available as sensible heat, resulting in higher land surface temperatures [1]. There is a strong relationship between impervious surface cover and land surface temperature in cities [2], with streets being especially susceptible [3]. Likewise, vegetated areas in cities are known to reduce maximum temperature, for example [4], [5], and urban designers are increasingly calling for more greenspace to mitigate UHI effects [6].

Previous research has shown that evapotranspiration (ET) from urban vegetation can have a significant cooling effect in cities. For example, a large greenspace in London caused temperature reductions of 1.1°C on summer months, and a maximum of 4°C observed on some nights, with cooling effects extending from 20–440 m from the edge of the greenspace [5]. Trees and shrubs reduced soil temperature by 5.7°C in a mid-sized city in the UK, which was 0.6°C warmer than nearby rural areas [7]. In Beirut, areas of the city with larger garden fractions have up to 6°C cooler temperatures compared to surrounding areas [8]. In addition to natural vegetation within cities, green roofs have been a focus of research on UHI mitigation [1], [9]. Green roofs cool individual buildings through a combination of ET, physical shading, and insulation, and also contribute to urban cooling [9], [10]. For

---

* ORCID: http://orcid.org/0000-0002-9018-7555

example, a model simulating 50% green roof coverage in a city predicted temperature reductions of 2°C [9].

In addition to natural vegetation in cities and highly engineered green roofs, urban vegetable gardens represent another land use that has received somewhat less attention for UHI mitigation potential. Urban gardens may be similar or more extensive in area compared to green roofs; urban vegetable gardens have been estimated to account for 0.04% of land area in Chicago [11], 0.1% of land area in Saint Paul, Minnesota [12], and 3.6% of land area in Montreal [13]. For comparison, Chicago, a leading city in green roof technology, has approximately 0.5 km^2 of vegetated square roofs in the 606 km^2 metropolitan area [14], or 0.08% land coverage. Additionally, unlike other forms of urban greenspace, many urban vegetable gardens receive supplemental irrigation during dry periods, increasing potential evapotranspiration and associated cooling. The potential impact of urban gardens on UHI mitigation has been considered in some previous analyses [1], [15], but empirical data on ET and cooling from urban gardens are limited.

We analyzed data from summer months over two years from a large urban garden in Saint Paul, Minnesota, USA, to evaluate the extent of UHI mitigation. We hypothesize that elevated ET rates from the irrigated garden will result in cooling rates that exceed documented rates for natural urban vegetation and green roofs.

2 METHODS

2.1 Garden description

The research garden at the University of Saint Thomas (Saint Paul, Minnesota, USA) is located at 44°56'17"N, 93°11'46"W. The garden, established in 2011, consists of 32 replicated 4 m^2 research plots growing peppers, carrots, bush beans, and collard greens. Research plots receive annual inputs of compost (either manure compost or municipal compost, targeted either to crop nitrogen or phosphorus demand), synthetic fertilizer, or no inputs. Garden soil, on average, has 9.4% organic matter, compared to surrounding soil below turfgrass which has 5.1% organic matter. Additional plots are dedicated to crop production and pollinator habitat, with a total area of approximately 1000 m^2. The garden is situated on a 1.8 ha park-like area with approximately 50% tree cover. The research garden is located 20 m from an adjacent classroom building, and 250 m from the Mississippi River.

2.2 Temperature measurements

Hourly temperature readings for the garden were collected using Decagon instruments/datalogger, installed at a height of 1.0 m near the center of the garden. To quantify potential UHI mitigation by the garden, we compared temperature data collected at the Minneapolis-Saint Paul International Airport (MSP), located 6.6 km SSW of the garden. MSP data were taken from the Iowa State climate data repository (http://mesonet.agron.iastate.edu). Temperature measurements at MSP are made in an open, grassy area located 35 m from an airplane taxiway, 250 m from the nearest building, 3.2 km from the Minnesota River, 3.5 km from the Mississippi River, and 2.1 km from Lake Nokomis.

Paired hourly temperature observations from the research garden and MSP airport were compiled from 1 June–30 August for 2017 and 2018, and any hours with missing observations were not used. 2,144 paired observations (out of a possible 2,184) were used for 2017, and 2,180 paired observations were used for 2018.

Regression analysis (using JMP Pro 15) was used to determine the slope of the relationship between airport temperature observations and garden temperature observations.

2.3 Evapotranspiration measurements

Evapotranspiration was calculated for the 2017 and 2018 growing seasons (30 May 2017–18 October 2017; 24 May 2018–20 October 2018), as difference between water inputs (rainfall and supplemental watering) and water loss through leachate. Rainfall was measured using an ECRN-50 rain gauge with 1 mm resolution, connected to a EM50 data logger (Decagon Devices). Supplemental watering of garden was quantified by applying water for a set amount of time (typically 30 or 45 seconds) to each 4 m^2 plot, and measuring the amount of time required to fill a 10 L bucket on each watering date. Water was distributed evenly across study plots. Turfgrass was not watered and received only ambient rainfall.

Leachate was quantified by weekly measurements of water volume collected in 128 lysimeters installed below each 1 m^2 garden subplot, and five additional lysimeters installed below turfgrass around perimeter of research garden. Each lysimeter consisted of a funnel with diameter 11.8 cm (surface area 109.3 cm^2) attached to a 1 L plastic bottle. The top of the funnel was buried to a depth of 10 cm (below the rooting zone). Flexible polymer tubing ran from the collection bottle to the soil surface, and stone wool was used to plug the hole in the funnel to prevent soil from entering the collection bottle. Water was removed from each lysimeter using a syringe weekly during the growing season (June–October), and the volume was measured using graduated cylinder. Lysimeters were replaced each growing season. Soil moisture was measured in all garden study plots and control turfgrass plots approximately three times per week during the growing season in 2017, using a General DSMM500 soil moisture meter.

3 RESULTS

3.1 Temperature measurements

The mean air temperature observed at the research garden during June–August 2017 was 22.6°C, compared to 23.3°C for MSP airport. Airport temperature exceeded garden temperature in 1,286 paired observations in 2017, whereas garden temperature was greater in 855 observations. For summer 2018, mean air temperature in the research garden was 21.4°C, compared to 22.1°C at the MSP airport. Airport temperature exceeded garden temperature in 1301 paired observations in 2018, whereas garden temperature was greater in 876 observations.

Regression analysis shows that an increase of 1.00°C at the MSP airport resulted in an average increase of 0.56 ± 0.02°C (mean ± SE) in the research garden in 2017 (Fig. 1), and 0.55 ± 0.02°C in 2018 (Fig. 2). Garden temperatures tended to be cooler than airport temperatures when airport temperatures exceeded 21°C, and warmer than airport temperatures at temperatures below 21°C during the summer months.

3.2 Evapotranspiration measurements

During the 2017 growing season, the UST research garden received 48.1 cm ambient rainfall, and 15.2 cm supplemental watering. During the 2018 growing season, the garden received 62.1 cm ambient rainfall, and 12.6 cm supplemental watering. Despite receiving 20–30%

2017

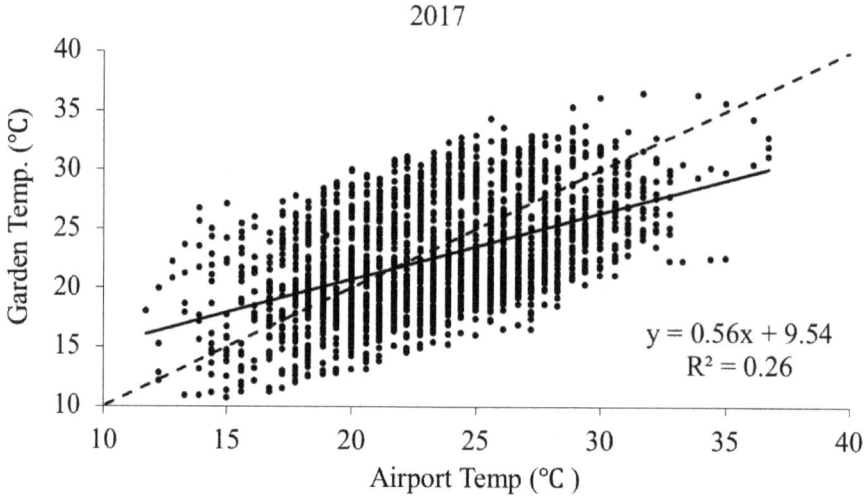

Figure 1:   Relationship between measured hourly air temperature at the UST research garden and MSP airport during summer 2017. The solid line is a regression line; the dashed line is a 1:1 line.

2018

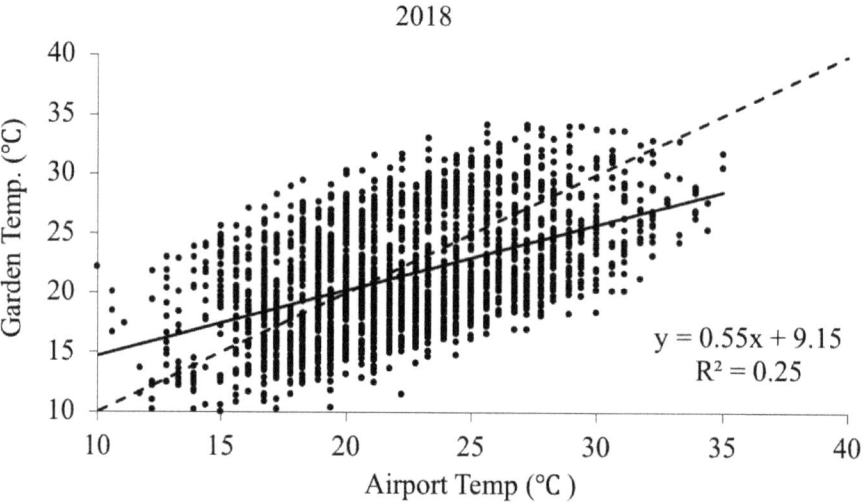

Figure 2:   Relationship between measured hourly air temperature at the UST research garden and MSP airport during summer 2018. The solid line is a regression line; the dashed line is a 1:1 line.

higher water inputs due to supplemental watering, garden plots had leachate fluxes 20–40% lower compared to grass plots (Table 1). Estimated ET flux in garden plots was 140% greater than in grass plots in 2017, and 55% greater in 2018 (Table 1). Soil moisture observations had a mean of 16.6% (± 0.1) in garden plots in 2017, compared to 13.4% (± 1.4) in grass plots.

Table 1:   Water budgets for garden plots and grass plots from 2017 (30 May–18 October) and 2018 (24 May–20 October) growing seasons.

	Garden plots n = 128 mean	SE	Grass plots n = 5 mean	SE
2017				
Total rainfall (cm)	48.1		48.1	
Supplemental watering (cm)	15.2		0.0	
Total water input (cm)	63.3		48.1	
Leachate (cm)	16.8	1.0	28.6	4.8
Evapotranspiration (cm)	46.5	1.0	19.5	4.8
2018				
Total rainfall (cm)	62.1		62.1	
Supplemental watering (cm)	12.6		0.0	
Total water input (cm)	74.0		62.1	
Leachate (cm)	23.7	0.9	29.2	6.3
Evapotranspiration (cm)	51.0	0.9	32.9	6.3

4 DISCUSSION

Our results support the hypothesis that elevated ET in an urban vegetable garden can contribute significantly to UHI mitigation, similar to cooling effects documented from green roofs and natural urban vegetation. Temperatures in the garden were moderated relative to airport temperatures, with a greater magnitude of cooling observed at higher temperatures. At temperatures above 21.7°C (in 2017) or 20.3°C (in 2018), garden temperatures tended to be cooler than temperatures at the MSP airport. For the warmest temperatures observed at MSP airport (approximately 35°C), garden temperatures were, on average, nearly 6°C cooler.

Our results are consistent with the few other available empirical data on cooling effects of urban gardens. A study in Rosario, Argentina, found that the cooling effects of urban agriculture gardens of only 0.2 ha were similar to the effects of gardens and parks that were 2–3 ha [16]. Likewise, the empirical data from our study is in agreement with results from a modeling study indicating that urban gardens can reduce surface temperatures by 5–10°C [15].

Evapotranspiration in garden plots was approximately double the rates observed in nearby turfgrass plots, potentially accounting for much of the observed cooling. While supplemental watering in the garden allowed for additional ET in garden plots, it is notable that cumulative leachate in garden plots was lower compared to turfgrass plots during both seasons, despite the additional water added. As a result, the total fraction of water inputs that are removed through ET is much higher in garden plots (73.4% in 2017 and 68.9% in 2018) relative to turfgrass control plots (40.5% in 2017 and 53.0% in 2018). The higher soil moisture observed in garden plots attests to the capacity of compost-amended garden soil to store water until it is used by plants. Many garden vegetable crops also have deeper root systems compared to turfgrass, allowing water to be extracted from deeper in the soil. Most leachate follows heavy rain events, when the capacity of the soil to retain water has been saturated. Because supplemental watering only occurs during relatively dry periods, the contribution to leaching is minimal; thus, ET is the fate of nearly all of this additional water. This finding is in agreement with the conclusions of Mariani et al. [17] which found that soil water reservoirs are important in UHI mitigation in urban parks. A synthesis of UHI data from 35 cities worldwide found that the effects of vegetative cooling increase sharply when ET exceeds 61.7 mm/month [18]; ET in our garden plots exceeded 90 mm/month, compared to rates of 40–60 mm/month for the turfgrass plots.

We cannot be sure that all cooling from the garden site was attributed to elevated ET in the garden itself; other factors, such as nearby tree cover or proximity to the Mississippi River may have also contributed to cooling. One study found cooling effects extending up to 300m from a large river [19], but other studies have found more limited [20] or no [21] cooling effects of rivers. While our study has only two spatial data points, a spatially-intensive study in the Minneapolis-Saint Paul Metropolitan Area using 170 sensors shows that both the UST research garden and the MSP airport are well within the UHI zone [22], experiencing summer daytime mean temperatures approximately 1.0°C higher than surrounding rural areas. That study showed no discernable cooling effect of proximity to the Mississippi or Minnesota Rivers [22].

Our study provides some of the first empirical evidence showing that high ET from urban gardens dedicated to vegetable production may result in local UHI mitigation. Because green roofs and other urban green space generally do not receive supplemental irrigation, ET from vegetable gardens is likely to be greater, potentially magnifying the local cooling effect. Moreover, since urban vegetable gardens are more widespread and less costly than green roofs, their cumulative effect on UHI mitigation may be substantial, and deserves more attention. Importantly, there are many other ecosystem services documented from outdoor urban crop production [23], [24], so UHI mitigation is an added benefit which may contribute to more livable, sustainable cities.

ACKNOWLEDGEMENTS

This study was supported in part by a National Science Foundation CAREER award (award number 1651361) to G.E. Small. J. Abrahamson, H. Dare, K. Dennis, A. Gilmore, M. Hay, A. Johnson, W. Kreuser, E. Mahre, C. Monroe, G. Pahl, T. Schmitt, I. Tjokrosetio, and S. Wihlm assisted with data collection.

REFERENCES

[1]   Qui, G., Li, H., Zhang, Q., Chen, W., Liang, X. & Li, X., Effects of evapotranspiration on mitigation of urban temperature by vegetation and urban agriculture. *Journal of Integrative Agriculture*, **12**, pp. 1307–1315, 2013.

[2]   Yuan, F. & Bauer, M.E., Comparison of impervious surface area and normalized difference vegetation index as indicators of surface urban heat island effects in Landsat imagery. *Remote Sensing of Environment Volume*, **106**, pp. 375–386, 2007.

[3]   Sharifi, E. & Lehmann, S., Comparative analysis of surface urban heat island effect in central Sydney. *Journal of Sustainable Development*, 7, pp. 23–24, 2014.

[4]   Wong, N.H. & Yu, C., Study of green areas and urban heat island in a tropical city. *Habitat International*, **29**, pp. 547–558, 2005.

[5]   Doick, K.J., Peace, A. & Hutchings, T.R., The role of one large greenspace in mitigating London's nocturnal urban heat island. *Science of the Total Environment*, **493**, pp. 662–671, 2014.

[6]   Icaza, L.E. & Van der Hoeven, F., Regionalist principles to reduce the urban heat island effect. *Sustainability*, **9**, p. 677, 2017.

[7]   Edmondson, J.L., Scott, I., Davies, G., Gaston, K.J. & Leake, J.R., Soil surface temperatures reveal moderation of the urban heat island effect by trees and shrubs. *Scientific Reports*, **6**, 33708, 2016.

[8]   Kaloustian, N. & Diab, Y., Effects of urbanization on the urban heat island in Beruit. *Urban Climate*, **14**, pp. 154–165, 2015.

[9]   Oberndorfer, E. et al., Green roofs as urban ecosystems: Ecological structures, functions, and services. *BioScience*, **57**, pp. 823–833, 2007.

[10] Lee, J.S., Kim, J.T. & Lee, M.G., Mitigation of urban heat island effect and greenroofs. *Indoor and Built Environment*, **23**, pp. 62–69, 2013.

[11] Taylor, J.R. & Lovell, S.T., Urban home food gardens in the global north: Research traditions and future directions. *Agriculture and Human Values*, **31**, pp. 85–305, 2014.

[12] Small, G., Shrestha, P., Metson, G.S., Polsky, K., Jimenez, I. & Kay, A., Excess phosphorus from compost applications in urban gardens creates potential pollution hotspots. *Environmental Research Communications*, **1**, 091007, 2019.

[13] Metson, G.S. & Bennett, E.M., Phosphorus cycling in Montreal's food and urban agricultural systems. *PLoS One*, **10**, e0120726, 2015.

[14] City of Chicago, Chicago green roofs, 2020. www.chicago.gov/city/en/depts/dcd/supp_info/chicago_green_roofs.html. Accessed on: 13 Jan. 2020.

[15] Tsilini, V., Papantoniou, S., Kolokots, D.-D. & Maria, E.-A., Urban gardens as a solution to energy poverty and urban heat island. *Sustainable Cities and Society*, **14**, pp. 323–333, 2015.

[16] Coronel, A.S. et al., Effects of urban green areas on air temperature in a medium-sized Argentinian city. *AIMS Environmental Science*, **2**, pp. 803–825, 2015.

[17] Mariani, L., Parisi, S.G., Cola, G., Lafortezza, R., Colangelo, G. & Sanesi, G., Climatological analysis of the mitigating effects of vegetation on the urban heat island of Milan, Italy. *Science of the Total Environment*, **569**, pp. 762–773, 2016.

[18] Su, Y. et al., Phenology acts as a primary control of urban vegetation cooling and warming: A synthetic analysis of global site observations. *Agricultural and Forest Meteorology*, **280**, 107765, 2020.

[19] Murakawa, S., Sekine, T., Narita, K.I. & Nishina, D., Study of the effects of a river on the thermal environment in an urban area. *Energy and Buildings*, **16**, pp. 993–1001, 1991.

[20] Hathway, E.A. & Sharples, S., The interaction of rivers and urban form in mitigating the urban heat island effect: A UK case study. *Building and Environment*, **58**, pp. 14–22, 2012.

[21] Moyer, A. & Hawkins, T.W., River effects on the heat island of a small urban area. *Urban Climate*, **21**, pp. 262–277, 2017.

[22] Smoliak, B.V., Snyder, P.K., Twine, T.E., Mykleby, P.M. & Hertel, W.F., Dense network observations of the Twin Cities canopy-layer urban heat island. *Journal of Applied Meterology and Climatology*, **54**, pp. 1899–1917, 2015.

[23] Ackerman, K., Dahlgren, E. & Xue, X., *Sustainable Urban Agriculture: Confirming Viable Scenarios for Production*, New York State Energy Research and Development Authority: Albany, NY, 2013.

[24] Nogeire-McRae, T. et al., The role of urban agriculture in a secure, healthy, and sustainable food system. *BioScience*, **68**, pp. 748–759, 2018.

WINTERTIME THERMAL PERFORMANCE OF GREEN FAÇADES IN A MEDITERRANEAN CLIMATE

ILEANA BLANCO[1*], FABIANA CONVERTINO[2†], EVELIA SCHETTINI[2‡] & GIULIANO VOX[2§]
[1]Department of Biological and Environmental Sciences and Technologies DiSTeBA, University of Salento, Italy
[2]Department of Agricultural and Environmental Science DISAAT, University of Bari, Italy

ABSTRACT

The increasing environmental issues have afforded opportunities for a widespread application of green systems in urban areas. Greening the building with green roofs and vertical green systems can be a design and retrofitting strategy to improve building energy performance in summer and in winter. Research efforts have been mainly concentrated on their energy saving function during warm periods. Green façades have a great application potential thanks to the space available in urban environment. The effect of green façades on building energy performance has been studied mainly for warm periods. In order to evaluate the effect during cold periods, an experiment was conducted in Bari, Italy, for two years. *Pandorea jasminoides variegated* and *Rhyncospermum jasminoides* were tested as evergreen climbing plants on walls; a third wall was used as control. The night-time temperature of the covered wall was higher than the uncovered wall temperature by up to 3.5°C, thanks to the presence of plants. The thermal barrier function performed by the vegetation layer was analysed. The influence of outdoor air temperature, relative humidity and wind velocity on the façades thermal effect during night-time was investigated. The experimental test demonstrated that both *Pandorea jasminoides variegated* and *Rhyncospermum jasminoides* are suitable for green façades in the Mediterranean climatic area during winter. The use of the green façades allowed increasing the thermal performance of the walls during night-time. They also reduced the surface temperature changes throughout the day.
Keywords: building energy efficiency, energy saving, energy balance, heating effect, vertical greenery systems.

1 INTRODUCTION

The European Council set the goal for Europe and other developed economies of cutting greenhouse gas emissions (GHG) by 80–95% by 2050 below 1990 levels [1]. EU countries succeeded between 1990 and 2017 in decoupling GHG emissions from economic growth by reducing GHG emissions by 22% and at the same time increasing gross domestic product by 58% [2]. Nevertheless, the decarbonisation process is still slow. The reduction of the consumption and environmental impacts of the building sector plays an important role for this objective. One third of global greenhouse gas emissions are attributable to buildings and the heating and cooling buildings energy demands account to about 50% of the final energy consumption in the EU28. Large energy savings can be obtained by improving the building envelope design and construction that affect 20–60% of all energy used in buildings to maintain internal thermal comfort [3], [4].

The ever-rising urban population leads to the replacement of natural vegetation with reinforced concrete buildings and thus to the urban heat island effect, the increase in building energy consumption and GHG emissions. Urban green infrastructures (UGI) are nature-based solutions that can improve urban climate conditions, decrease urban air, surfaces temperature

---

* ORCID: *http://orcid.org/0000-0003-2927-3427*
† ORCID: *http://orcid.org/0000-0002-3799-6038*
‡ ORCID: *http://orcid.org/0000-0002-8456-8677*
§ ORCID: *http://orcid.org/0000-0003-4017-5174*

WIT Transactions on Ecology and the Environment, Vol 243, © 2020 WIT Press
www.witpress.com, ISSN 1743-3541 (on-line)
doi:10.2495/UA200051

level and variation, in particular in the Mediterranean climatic area [5]. UGI include lawns, parks, private gardens, shade trees, remaining native vegetation, and so on, and building greenery systems such as green balconies, sky gardens, green roofs, green walls [6], [7]. The most studied building greenery systems are the green roofs, nevertheless green walls are gaining even more attention. Green walls have potentially more environmental effectiveness in highly urbanized areas because the building vertical surfaces can be also 20 times the area of the roof [8], [9]. Moreover, a very little space for greening results on the rooftop due to the presence of bulky devices, such as solar panels and water tanks.

Green wall systems can be classified into green façades and living walls. Green façades are characterized by climbing plants that grow on a building vertical wall attached directly to it (direct green façades) or on a structural support such as modular trellis, wire, mesh (indirect green façades). The supporting structure is located to a small distance from the wall. Climbers can be rooted in the soil on ground or in pots that can be placed at different heights of the façade. Living walls are characterized by the growing media embedded in panels, modules, bags (modular living walls) or by continuous screen or geotextile felt (continuous living walls) which are fixed to a wall or a free-standing frame [6], [10]. Green façades, when compared to living walls have a wider application potential due to having a simpler composition, easier installation and maintenance, lower operational and installation costs [11].

The use of vertical greenery systems is a passive green technology for improving buildings energy performance by lowering energy demand for air conditioning in summer and by enhancing thermal insulation in winter. Their application can also alleviate urban ecological environment deterioration and provides economic and social benefits [12]–[16]. Thermal effects of green façades on the environment are mainly due to the following mechanisms: shading of the building envelope from solar radiation provided by plants, thermal insulation produced by the different layers composing the greenery system, cooling by evapotranspiration from the vegetation or from the substrate, and protection from wind exposure [15]–[17]. Moreover, the creation of an air gap between the building external wall and the green layer can act as a thermal barrier to the diffusion of longwave radiation from the wall towards external ambient, improving building thermal insulation [6], [8], [18].

Many experimental and theoretical studies have mainly investigated thermal effects of green façades on buildings in summer. A few studies have investigated their energy behaviour throughout the year for evaluating the performance of perennial climbing plants [8], [19].

Winter passive warming may come from the higher temperatures generated in the air-gap and on the wall external surface in indirect green façades [20]. In subtropical climates, where buildings are rarely equipped with air heating systems, a negative temperature difference could be generated by keeping warmth in the air gap and wall external surface, thus a heat flux from them towards indoor ambient could be established [21].

Highest thermal performances of green façades have been assessed by Cameron et al. [22] during extreme weather conditions, i.e. very low temperatures, high wind or heavy rain in temperate oceanic climate. Energy savings up to 50% and wall surface temperatures higher up to 3°C, compared with a bare wall, were recorded using a Hedera helix as green façade component.

Wall temperatures and energy loss on a Hedera helix green façade and on a bare exposed wall, both north-facing, were compared in a maritime temperate climate. Minimum external wall temperatures on the green façade were on average 1.7°C higher than the bare wall, winter heating costs were reduced, and energy losses were lowered by almost 8%, despite the heating effect of short-wave radiation was minimized during daytime [23].

In winter the vegetated systems could create negative thermal effects by reducing the inflow of short-wave solar radiation and thus, their application could lead to an increase in heating systems energy consumptions [24].

Green façades thermal benefits in winter conditions should be assessed and plant species suitable for this application in areas characterized by high levels of solar radiation, such as the Mediterranean regions, should be defined [25].

This study aims to investigating the thermal performance of two different evergreen climbing plants for green façades, experimentally tested at the University of Bari in the winter season. Climatic data and surface temperatures were analysed for defining the influence of the different climatic parameters on the external surface temperatures of the walls equipped with vertical greenery systems. Benefits deriving from the thermal barrier effect provided by the vegetation layer were also evaluated.

2 MATERIALS AND METHODS

A research was performed at the experimental centre of the University of Bari in Valenzano (Bari, Italy) latitude 41° 01' N and longitude 16° 54' E, in the period June 2014–December 2016. The climate at the experimental field is classified as Csa, Mediterranean climate, according to the Köppen–Geiger climate classification [26]. It is a warm temperate climate with a particularly evident variation of solar radiation intensity at seasonal level; the average annual temperature is 16.1°C and the winter months are considerably rainier than the summer months.

A typical Mediterranean building solution was followed for the setup of three experimental walls in open environment, built facing south. The walls are made of perforated bricks, arranged in a single skin, joined with mortar (Fig. 1) with a width of 1.00 m, a height of 1.55 m, and a total thickness of 0.22 m, including 0.02 m of the plaster coating. The masonry is characterized by a thermal conductivity coefficient of 0.28 $Wm^{-1}K^{-1}$ [27] and a heat capacity coefficient of 840 $Jkg^{-1}K^{-1}$. The plaster coating is characterized by a thermal

Figure 1: The three walls at the experimental field of the University of Bari: the uncovered control (the left wall), the *Pandorea jasminoides variegated* green façade (on the central wall), the *Rhyncospermum jasminoides* green façade (on the right wall).

conductivity coefficient of 0.55 Wm^{-1}K^{-1} and a heat capacity coefficient of 1,000 Jkg^{-1}K^{-1}. The average density of the walls, taking in consideration also plaster, was equal to 695 kg m^{-3}. The south exposed side of two walls was covered with plants and it simulates the external side of a Mediterranean common building envelope. The other side was shaded and insulated with a sealed structure made of sheets of expanded polystyrene. The incident solar radiation effect on the sealed structure was reduced by the adoption of a shading net, positioned onto the structures.

The radiometric properties of the wall surface were evaluated by laboratory tests [28]. The emissivity coefficient of the wall surface samples in the long wave infrared radiation (LWIR) range (2,500–25,000 nm) was 95.3% and the solar absorption coefficient was 42.1%.

Pandorea jasminoides variegated and *Rhyncospermum jasminoides* were chosen for their capacity to easily climb the wall and to grow vigorously in the climatic conditions of the experimental area. The plants were transplanted on June 18, 2014. The third bare wall was kept for control. As structure supporting plant growing, an iron net was put 15 cm away from the wall. Drip irrigation and fertilization with N:P:K 12:12:12 were performed. Plant leaf surface index (LAI) varied throughout the year in the ranges 1.5–3.5 and 2.0–4.0 for *Pandorea jasminoides variegated* and *Rhyncospermum jasminoides*, respectively. It was measured with an AccuPAR PAR/LAI Ceptometer (model LP-80, Decagon Devices Inc., Pullman, WA, USA).

The external air temperature and relative humidity, the wind speed and direction, the external surface temperature of the wall, the solar radiation on a horizontal and on a vertical plane were measured during the experimental test. The data were measured with a frequency of 60 s averaged every 15 min and recorded on a data logger (CR10X, Campbell, Logan, USA). The external air temperature was measured by a Hygroclip-S3 sensor (Rotronic, Zurich, Switzerland), adequately shielded from solar radiation. The temperature on the external surface of the walls was measured using thermistors (Tecno.EL s.r.l. Formello, Rome, Italy). The solar radiation on a horizontal and on a vertical plane were measured by means of pyranometers (model 8-48, Eppley Laboratory, Newport, RI, USA) in the wavelength range 300–3000 nm.

Analysis of variance (ANOVA), in detail a three-way ANOVA analysis at 95% probability level, was performed for assessing the influence of external air temperature (EAT), wind speed (W) and air relative humidity (RH) of the outdoor environment on the heating effect of the façades. The CoStat software (CoHort Software, Monterey, CA, USA) was used to carry out the ANOVA.

The thermal barrier effect of the vegetation layer and the deriving benefits, in the cold season, were also analysed. The longwave infrared energy balance at the external surface of the bare wall and of the wall covered with *Rhyncospermum jasminoides* was evaluated. Calculations were performed for a representative winter day, according to Convertino et al. [29].

For the external surface of the bare wall, the LWIR radiative balance (RB) is equal to:

$$RB_{bw} = \varepsilon_{ws}\left(R_{sky} + R_g\right) - R_{e,bw}, \tag{1}$$

where: ε_{ws} is the infrared emissivity coefficient of the wall external surface, R_{sky}, R_g and $R_{e,bw}$ (Wm^{-2}) are the LWIR radiative fluxes emitted by the sky, the ground and the external surface of the bare wall, respectively.

RB for the external surface of the covered wall is:

$$RB_{cw} = \varepsilon_{ws}R_{i,gl} - R_{e,cw}, \tag{2}$$

where $R_{i,gl}$ and $R_{e,cw}$ (Wm^{-2}) are the LWIR radiative fluxes emitted by the inner side of the green layer of vegetation and by the external surface of the covered wall, respectively.

3 RESULTS AND DISCUSSION

Air temperatures and monthly values of cumulative solar radiation on the horizontal and vertical planes corresponding to 2015 and 2016 winter months are summarized in Table 1. The minimum and maximum monthly value of cumulative solar radiation on a horizontal plane were recorded respectively in December and February, in 2015, and in January and February, in 2016. The minimum and maximum monthly value of cumulative solar radiation on the vertical wall were recorded respectively in February and January, in 2015, and in January and December, in 2016.

Table 1: Air temperatures and monthly values of cumulative solar radiation on a horizontal and on a vertical plane on the experimental field of the University of Bari in 2015 and 2016 winter months (January, February and December).

	Air temperatures (°C)			Monthly cumulative solar radiation on a horizontal plane (MJ m^{-2})		Monthly cumulative solar radiation on a vertical plane (MJ m^{-2})	
	Mean	Minimum	Maximum	Minimum	Maximum	Minimum	Maximum
2015	9.6	−0.3	20.9	179	213	254	338
2016	10.9	0.7	22.6	177	238	289	333

The climatic conditions able to mostly influence during night-time the heating effect of the two green façades were examined. The heating effect was assumed as the positive difference between the external surface temperature of the wall behind the vegetation and the temperature of the external surface of the bare wall.

The maximum and average heating effect obtained in the analysed period were equal to 3.5°C and 1.1°C for *Rhyncospermum jasminoides* and to 3.5°C and 1.2°C for *Pandorea jasminoides*, respectively. The maximum cooling effect during daytime was of 8.3°C and 7.7°C for *Rhyncospermum jasminoides* and *Pandorea jasminoides*, respectively. These findings are confirmed by results available in the literature that refer to slightly different climates [22], [23].

The ANOVA statistical analysis revealed significant differences at P<0.001 (Tables 2 and 3). Subsequently, the effects of the three external climate factors were analysed by means of the Tukey–Kramer's test (Tables 4–6).

It emerges that the variability determined by EAT is greater than those determined singly both by W and by RH, and by the interaction of the climatic parameters, for both *Rhyncospermum jasminoides* and *Pandorea jasminoides* façades.

The Tukey–Kramer's test highlighted the dependence of the magnitude of the heating effect on EAT for both the two green façades (Table 4). The heating effect is more evident when air temperature decreases. Moreover, the analysis showed that for EAT values ≥ 10°C *Rhyncospermum jasminoides* and *Pandorea jasminoides* façades have almost a steady effect on the external surface temperature of the walls. W affected the heating performance with an increasing trend for speed values lower than 3 ms^{-1}; once exceeded this limit the heating effect varies with a slightly decreasing trend. The maximum effect occurs in the range 2–3 ms^{-1} (Table 5). The Tukey–Kramer's test shows the scarce influence of RH on the

Table 2: ANOVA results concerning the external climate conditions influence on the night-time temperature rise of the wall surface behind the *Rhyncospermum jasminoides*.

Source[a]	df	MS	F	P
Main effects				
EAT	7	103.96	398.41	***
W	5	23.65	90.65	***
RH	6	2.30	8.82	***
Interaction				
EAT × W	35	4.42	16.95	***
EAT × RH	37	2.87	10.99	***
W × RH	28	1.01	3.88	***
EAT × W × RH	131	0.82	3.16	***
Error	10,722	0.26		

[a] EAT: external air temperature; W: wind speed; RH: relative humidity.
*** $P \leq 0.001$.

Table 3: ANOVA results concerning the external climate influence on the night-time temperature rise of the wall surface behind the *Pandorea jasminoides*.

Source [a]	df	MS	F	P
Main effects				
EAT	7	59.89	218.84	***
W	5	14.67	53.60	***
RH	6	8.38	30.63	***
Interaction				
EAT × W	35	3.82	13.97	***
EAT × RH	37	3.84	14.02	***
W × RH	28	1.19	4.34	***
EAT × W × RH	131	0.98	3.59	***
Error	10,640	0.27		

[a] EAT: external air temperature; W: wind speed; RH: relative humidity.
*** $P \leq 0.001$.

heating effect. The maximum effect of RH on the heating effect has been achieved for percentage values falling in the range 70–80% for the *Pandorea jasminoides* façade (Table 6). In this façade the heating performance was influenced by RH with an increasing trend from 40 to 80%; once exceeded this limit the heating effect starts decreasing.

The calculation of the LWIR radiative energy budget at the external surface of the bare wall and of the wall behind *Rhyncospermum jasminoides* brought out significant differences in the thermal behaviour of the two solutions (Fig. 2).

The LWIR radiative fluxes were evaluated distinguishing between daytime, night-time and all day (Fig. 2), since they are strongly related to the day period.

During daytime, the bare wall lost a quantity of LWIR energy equal to 1.23 MJm^{-2}, while the covered wall gains 0.05 MJm^{-2}. At night-time, the LWIR energy lost by the bare wall (2.32 MJm^{-2}) was 56% higher than that lost by the covered wall (1.01 MJm^{-2}).

Overall, daily, the bare wall lost 73% more energy than the covered wall (Fig. 2).

Table 4: Mean temperature rise of the wall external surface, during night-time, as related to the external air temperature levels, analysed with Tukey–Kramer's test.

		EAT < 4	$4 \leq$ EAT < 6	$6 \leq$ EAT < 8	$8 \leq$ EAT < 10	$10 \leq$ EAT < 12	$12 \leq$ EAT < 14	$14 \leq$ EAT < 16	EAT ≥ 16
Mean temperature rise (°C)	Rhyncospermum jasminoides	1.62[a]	1.48[b]	1.34[c]	1.10[d]	0.88[e]	0.87[ef]	0.84[ef]	0.79[f]
	Pandorea jasminoides	1.63[a]	1.45[b]	1.35[c]	1.20[d]	1.05[e]	0.95[e]	0.94[e]	0.91[e]

a-b-c-d-e-ef-f Mean values of the temperature in a row with a different superscript letter statistically differ at P < 0.05 using Tukey–Kramer's test.

Table 5: Mean temperature rise of the wall external surface, during night-time, as related to the wind speed levels, analysed with Tukey–Kramer's test.

		W < 2	$2 \leq$ W < 3	$3 \leq$ W < 4	$4 \leq$ W < 5	$5 \leq$ W < 6	W ≥ 6
Mean temperature rise (°C)	Rhyncospermum jasminoides	1.02[c]	1.33[a]	1.22[b]	0.99[cd]	0.90[de]	0.88[e]
	Pandorea jasminoides	1.18[c]	1.36[a]	1.25[b]	1.01[d]	0.95[de]	0.87[e]

a-b-c-cd-de-e Mean values of the temperature in a row with a different superscript letter statistically differ at P < 0.05 using Tukey–Kramer's test.

Table 6: Mean temperature rise of the wall external surface, during night-time, as related to the relative humidity levels, analysed with Tukey–Kramer's test.

		RH < 40	$40 \leq$ RH < 50	$50 \leq$ RH < 60	$60 \leq$ RH < 70	$70 \leq$ RH < 80	$80 \leq$ RH < 90	RH ≥ 90
Mean temperature rise (°C)	Rhyncospermum jasminoides	0.98[a]	1.15[a]	1.27[a]	1.16[a]	1.25[a]	1.20[a]	1.14[a]
	Pandorea jasminoides	0.90[c]	1.08[c]	1.21[bc]	1.28[b]	1.39[a]	1.28[b]	1.19[c]

a-b-bc-c Mean values of the temperature in a row with a different superscript letter statistically differ at P < 0.05 using Tukey–Kramer's test.

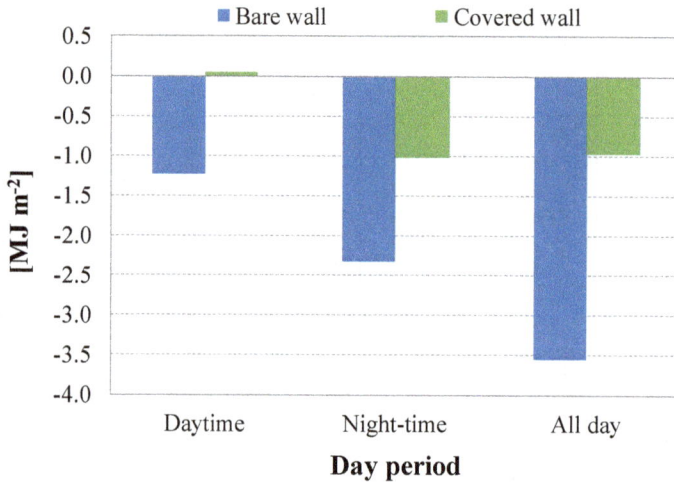

Figure 2:   Longwave radiative energy budget at the external surface of the bare wall and of the wall covered with *Rhyncospermum jasminoides*: daytime, night-time and all day.

These findings demonstrated the thermal barrier effect of the vegetation layer in winter. Indeed, the presence of the green façade reduced the energy losses caused by LWIR radiation.

4 CONCLUSIONS

The experimental test allowed quantifying and deepening the heating effect performed by the green layers during winter months of 2015–2016. The application of the vegetation allowed maintaining the temperature of the external facade of the walls covered with vegetation at higher values than the bare wall during night-time periods. A direct consequence is that the thermal losses, especially by longwave infrared radiation, decreased. It was found that the covered wall lost significantly less LWIR energy than the bare wall, during a representative winter day. The statistical analysis allowed understanding that the heating effect of the two green façades was mainly driven by the external air temperature with a greater increase as the temperature drops. Secondly, the heating effect magnitude resulted sensible to wind speed values, in particular to values in the range 2–3 ms^{-1}. Based on the climatic conditions, the findings of the research allow to define the sites where the green walls could have greater potential in terms of winter heating effect. Air relative humidity showed a scarce influence. Further future research should be addressed at quantifying the possible reduction of energy loss throughout the heating season.

ACKNOWLEDGEMENTS

The contribution to programming and conducting this research must be equally shared between the authors.

REFERENCES

[1]   European Commission, Energy Roadmap 2050. Publications Office of the European Union, 2012. https://ec.europa.eu/energy/sites/ener/files/documents/2012_energy_ roadmap_2050_en_0.pdf. Accessed on: 15 Jan. 2020. https://doi.org/10.2833/10759.

[2]   European Commission, Going climate-neutral by 2050: A strategic long-term vision for a prosperous, modern, competitive and climate-neutral EU economy, ML-04-19-339-EN-C. Publications Office of the European Union, 2019. https://op.europa.eu/en/publication-detail/-/publication/92f6d5bc-76bc-11e9-9f05-01aa75ed71a1/language-en/format-PDF/source-97097568. Accessed on: 15 Jan. 2020. https://doi.org/10.2834/508867.

[3]   Coma, J., Chàfer, M., Pérez, G. & Cabeza, L.F., How internal heat loads of buildings affect the effectiveness of vertical greenery systems? An experimental study. *Renewable Energy*, 2019. https://doi.org/10.1016/j.renene.2019.11.077.

[4]   De Masi, R.F., de Rossi, F., Ruggiero, S. & Vanoli, G.P., Numerical optimization for the design of living walls in the Mediterranean climate. *Energy Conversion and Management*, **195**, pp. 573–586, 2019. https://doi.org/10.1016/j.enconman.2019.05.043.

[5]   Schettini, E., Campiotti, C.A., Blanco, I. & Vox, G., Green façades to enhance climate control inside buildings. *Acta Horticulturae*, **1227**, pp. 77–84, 2018. https://doi.org/10.17660/ActaHortic.2018.1227.9.

[6]   Raji, B., Tenpierik, M.J. & van den Dobbelsteen, A., The impact of greening systems on building energy performance: A literature review. *Renewable and Sustainable Energy Reviews*, **45**, pp. 610–623, 2015. http://dx.doi.org/10.1016/j.rser.2015.02.011.

[7]   Blanco, I., Schettini, E., Scarascia Mugnozza, G., Campiotti, C.A., Giagnacovo, G. & Vox G., Vegetation as a passive system for enhancing building climate control. *Acta Horticulturae*, **1170**, pp. 555–561, 2017. https://doi.org/10.17660/ActaHortic.2017.1170.69.

[8]   Pérez, G., Coma, J., Martorell, I. & Cabeza, L.F., Vertical greenery systems (VGS) for energy saving in buildings: A review. Renewable and Sustainable Energy Reviews, 39, pp. 139–165, 2014. http://dx.doi.org/10.1016/j.rser.2014.07.055.

[9]   Blanco, I., Schettini, E., Mugnozza, G.S. & Vox, G., Thermal behaviour of green façades in summer. *Journal of Agricultural Engineering*, **49**(3), pp. 183–190, 2018. https://doi.org/10.4081/jae.2018.835.

[10]   Seyam, S., The impact of greenery systems on building energy: Systematic review. *Journal of Building Engineering*, **26**, 100887, 2019. https://doi.org/10.1016/j.jobe.2019.100887.

[11]   Zhang, L. et al., Thermal behavior of a vertical green facade and its impact on the indoor and outdoor thermal environment. Energy and Buildings, 204, 109502, 2019. https://doi.org/10.1016/j.enbuild.2019.109502.

[12]   Cheng, C.Y., Cheung, K.K.S. & Chu, L.M., Thermal performance of a vegetated cladding system on façade walls. *Building and Environment*, **45**(8), pp. 1779–1787, 2010. http://dx.doi.org/10.1016/j.buildenv.2010.02.005.

[13]   Köhler, M. & Poll, P.H., Long-term performance of selected old Berlin greenroofs in comparison to younger extensive greenroofs in Berlin. *Ecological Engineering*, **36**(5), pp. 722–729, 2010. http://dx.doi.org/10.1016/j.ecoleng.2009.12.019.

[14]   Manso, M. & Castro-Gomes, J., Green wall systems: A review of their characteristics. *Renewable and Sustainable Energy Reviews*, **41**, pp. 863–871, 2015. https://doi.org/10.1016/j.rser.2014.07.203

[15]   Perini, K., Ottelé, M., Fraaij, A.L.A., Haas, E.M. & Raiteri, R., Vertical greening systems and the effect on air flow and temperature on the building envelope. *Building and Environment*, **46** (11), pp. 2287–2294, 2011. http://dx.doi.org/10.1016/j.buildenv.2011.05.009.

[16] Santamouris, M., Cooling the cities: A review of reflective and green roof mitigation technologies to fight heat island and improve comfort in urban environments. *Solar Energy*, **110**, pp. 682–703, 2012. https://doi.org/10.1016/j.solener.2012.07.003.

[17] Convertino, F., Vox, G. & Schettini, E., Convective heat transfer in green façade system. *Biosystems Engineering*, **188**, pp. 67–81, 2019. https://doi.org/10.1016/j.biosystemseng.2019.10.006.

[18] Blanco, I., Schettini, E. & Vox, G., Effects of vertical green technology on building surface temperature. *International Journal of Design and Nature and Ecodynamics*, **13**(4), pp. 384–394, 2018. https://doi.org/10.2495/DNE-V13-N4-384-394.

[19] Blanco, I., Schettini, E. & Vox, G., Predictive model of surface temperature difference between green façades and uncovered wall in Mediterranean climatic area. *Applied Thermal Engineering*, **163**, 114406, 2019. https://doi.org/10.1016/j.applthermaleng.2019.114406.

[20] Schettini, E., Campiotti, C.A., Scarascia Mugnozza, G., Blanco, I. & Vox, G., Green walls for building microclimate control. *Acta Horticulturae*, **1215**, pp. 73–76, 2018. https://doi.org/10.17660/ActaHortic.2018.1215.13.

[21] Jim, C.Y., Cold-season solar input and ambivalent thermal behavior brought by climber greenwalls. *Energy*, **90**(1), pp. 926–938, 2015. https://doi.org/10.1016/j.energy.2015.07.127.

[22] Cameron, R.W.F., Taylor, J. & Emmett, M., A Hedera green façade: Energy performance and saving under different maritime-temperate, winter weather conditions. *Building and Environment*, **92**, pp. 111–121, 2015. https://doi.org/10.1016/j.buildenv.2015.04.011.

[23] Bolton, C., Rahman, M.A., Armson, D. & Ennos, A.R., Effectiveness of an ivy covering at insulating a building against the cold in Manchester, UK: A preliminary investigation. *Building and Environment*, **80**, pp. 32–35, 2014. https://doi.org/10.1016/j.buildenv.2014.05.020.

[24] Vox, G., Scarascia Mugnozza, G., Blanco, I. & Schettini, E., Heat fluxes in green walls. *Acta Horticulturae*, **1215**, pp. 273–278, 2018. https://doi.org/10.17660/ActaHortic.2018.1215.49.

[25] Campiotti, C.A. et al., Building green covering for a sustainable use of energy. *Journal of Agricultural Engineering*, **44**, pp. 253–256, 2013. http://dx.doi.org/10.4081/jae.2013.292.

[26] Kottek, M., Grieser, J., Beck, C., Rudolf, B. & Rubel, F., World map of the Köppen–Geiger climate classification updated. *Meteorologische Zeitschrift*, **15** (3), pp. 259–263, 2006. https://doi.org/10.1127/0941-2948/2006/0130.

[27] UNI, *UNI EN 1745. Masonry and masonry products: Methods for determining thermal properties*, Ente Nazionale Italiano di Unificazione: Milan, Italy, 2012.

[28] Vox, G., Blanco, I. & Schettini, E., Green façades to control wall surface temperature in buildings. *Building and Environment*, **129**, pp. 154–166, 2018. https://doi.org/10.1016/j.buildenv.2017.12.002.

[29] Convertino, F., Vox, G. & Schettini, E., Heat transfer mechanisms in vertical green systems and energy balance equations. *International Journal of Design and Nature and Ecodynamics*, **14**(1), pp. 7–18, 2019. https://doi.org/ 10.2495/DNE-V14-N1-7-18.

HEAT TRANSFER MODELLING IN GREEN FAÇADES

FABIANA CONVERTINO*
Department of Agricultural and Environmental Science, University of Bari, Italy

ABSTRACT

Making green urban environments through the application of the vertical green façades is an interesting new challenge in view of the growing sustainability of cities. Several advantages have been recognized for this type of passive technology. Inhabitants, buildings and cities are all recipients of these benefits. Due to their potential, an in-depth knowledge of the energy functioning of green façades and, consequently, their conscious use has become essential. This study traces a path for the energy analysis and modelling of green façade systems. Heat fluxes were identified and described by using a mathematical methodology and with the support of experimental data. The energy balance approach was followed. The most significant heat and mass transfer mechanisms were deepened and modelled. Convective, radiative and evapotranspirative exchanges were considered. Thanks to this strategy, the surface energy budget at the external building surface was evaluated. A comparison between the energy behaviour of a green façade and that of a bare wall was carried out. A measurement campaign relative to an experimental green façade realized at the University of Bari allowed also the implementation of realistic data. It was observed that the covered wall, behind the vegetation layer, gained 88% less energy than the bare wall during daytime in the summer. The findings demonstrated, with knowledge of the facts, that the green façades applied to buildings provide significant advantages in Mediterranean areas. In perspective, this study could be further developed through the writing of codes for the energy simulation of building equipped with green façades.
Keywords: green infrastructures, building envelope, vertical greening, energy balance, energy saving.

1 INTRODUCTION

The climatic and environmental emergency is now a fact and can no longer be ignored. All human activities, in every sector, must be oriented towards increasing sustainability. The constructive activities and the building sector have a great negative impact; thus, a sustainable turn can produce significant improvements.

Concerning this, a great opportunity is provided by the spread of the urban green infrastructures (UGIs) [1], [2]. The concept of UGI is wide and refers to several planned and unplanned solutions which bring vegetation inside cities [3]. Vegetation is, in fact, the key to solve numerous environmental problems and to improve the quality of urban areas.

A promising way, under many points of view, to integrate vegetation into urban contexts is its application on buildings surfaces [4]. These kinds of UGIs belong to two main groups of technological solutions: green roofs and vertical green systems (VGS) [5]. In the past few years, green roofs were the most studied and applied typology. Recently, VGSs received increasing research interest and their applications in cities multiplied. An indisputable advantage of using VGSs instead of green roofs is linked to the surfaces available in cities. The overall vertical surface is certainly wider than the horizontal one. However, concerning VGSs, other research contributions are needed to fill some gaps, especially to clarify their real performance and the most relevant influencing factors [6].

Many technical solutions are included in the expression VGSs. In any case, it indicates that a layer of vegetation covers the vertical walls of building. Plants can be directly attached to the wall or placed at a certain distance from it. They can be rooted into the ground or in pots and supported by metal structures (green façades) or in modular panels (green walls)

ORCID: http://orcid.org/0000-0002-3799-6038

WIT Transactions on Ecology and the Environment, Vol 243, © 2020 WIT Press
www.witpress.com, ISSN 1743-3541 (on-line)
doi:10.2495/UA200061

[7], [8]. Beyond the multiple available technical solutions, what needs to be highlighted is that VGSs are passive sustainable envelope technologies providing various kinds of positive contributions. The reduction of the urban heat island effect, the lowering of the carbon dioxide emissions and the rainwater retention are among the VGSs benefits at the urban scale. At the same time, VGSs enhance thermal performance of buildings, contributing to energy saving and human comfort. The application of greenery systems, as the recovery of the traditional bioclimatic architecture and the careful choice of constructive materials properties can contribute to the improvement of microclimatic conditions and human wellbeing [9]–[15].

This study focuses on a particular type of VGSs: the indirect green façade (GF). These GFs can be considered as double skin façades. In this case the second skin is the green layer of vegetation. This consists of evergreen or deciduous plants, climbing from below or cascading from above. Generally, a net is placed to assist the growth of the plants. This supporting structure allows to create an air gap between the building wall and the vegetation layer, since it stands at a certain distance from the building envelope. As other VGSs, GFs are a means of passive climate control in buildings. In fact, they are able to reduce the building heat transfer with the external environment. Many environmental and constructive factors influence the thermal performance of GFs: the climate, the building features and the plant characteristics are the most significant ones [16]–[18]. The positive contributions to the energy performance of the building envelope provided by the GF system can be summarized into three main functions. The green layer applied to building acts as a shading barrier during daytime [19]–[21]. At the same time, the vegetation, together with the air gap, increases the building envelope thermal insulation [22], [23]. A significant additional benefit is provided by the evapotranspiration of the plants and of the ground [24], [25]. As consequences, a building equipped with a GF has lower surface and air temperature and needs less energy for cooling, in summer [26]–[28]. In wintertime, the increased insulation and the effects of thermal and wind barrier provide the main benefits [29]–[31].

Evaluation of GFs performance mainly comes from empirical measurements on real scale buildings or small experimental blocks. In fact, a significant gap, concerning GFs, is the poor availability of energy models and simulation tools specific for them. Specific GFs tools implemented in building energy simulation programs would allow to simulate the energy behaviour of buildings equipped with GFs and, thus, to analyse them under different conditions and for long time periods. In this way, GFs would be designed in the best way and their energy functioning could be forecasted.

The design of simulation tools needs the energy model of a GF. The GF modelling implies that all the heat and mass transfer mechanisms occurring at each layer must be defined [32]–[35]. Heat fluxes due to radiation, convection and evapotranspiration are the main ones.

This paper focuses on the energy behaviour of the wall in the GF system, compared to a bare one. The influence of the presence or not of the green layer is clear by analysing the energy balance at the external surface of the two walls. In this research, radiative and convective energy transfer were mathematically defined and calculated for the two envelope solutions. The effect of the evapotranspiration, which is the most specific of a GF, can be read in the air characteristics near the GF system, which are different from those near the bare wall. Experimental data were gathered and used in calculations. The different energy behaviour of the walls was emphasized during summer. The surface energy budget was analysed with particular attention to the period of the day.

2 MATERIALS AND METHODS

Heat fluxes occurring at the external surface of a green façade system and of a bare wall, used as control, were studied. Three layers were identified to schematize these envelope solutions and the heat transfer mechanisms: the bare wall (BW), the green layer (GL) and the covered wall (CW), which stands behind the vegetation in the GF system.

Heat fluxes at the external surface of the BW and of the CW were evaluated. Radiative exchanges in the solar wavelength range and in the IR longwave range, and convective heat flow were determined.

Data collected during a measurement campaign concerning an experimental green façade were implemented in calculations.

The calculated heat fluxes at the external surface of these two technological solutions allow to highlight their different energy behaviours and the possible advantages deriving from the application of GFs in the Mediterranean climate.

2.1 Experimental set up

The experimental data used in the energy modelling have been collected at the experimental centre of the University of Bari, in Valenzano (Italy), where a measurement campaign is still in progress since 2018. The experimental set up, which simulates a real building equipped with a green façade, consists of a block, having a rectangular plane. Its south facing wall was studied. This was made of perforated bricks, held together with cement mortar, and completed with a white plaster external finishing. Such constructive solution was chosen to reproduce a typical external envelope used in the Mediterranean region. The south facing wall was divided into three parts: two parts provided with a green façade system and the last left bare as control (Fig. 1). The selected type of green façade was the indirect or double-skin one. The layer of vegetation was placed at 0.15 m from the external surface of the wall and, thus, an air gap was created between them. Plants of *Rhyncospermum Jasminoides*, an evergreen climbing species, were selected and an iron net was used to support their upward growth. Three plants were placed in pots, while other three were rooted directly into the ground.

Figure 1:   South facing wall and measurement instrumentation of the experimental set up. From right to left: green façade with plants rooted into the ground, green façade with plants in pots and bare wall (control).

The experimental equipment consists of a meteorological station, three data loggers (CR10X and CR1000, Campbell, Logan, USA) and sensors for climatic parameters detection. A pyranometer (model 8-48, Eppley Laboratory, Newport, RI, USA) allowed to measure solar radiation normal to the wall; wind speed and direction were detected using a Wind Sentry anemometer (model 03002, R. M. Young Company, USA). HygroClip-S3 sensors (Rotronic, Zurich, Switzerland) were used to measure air temperature and relative humidity; walls surface temperature was measured by thermistors (Tecno.el s.r.l. Formello, Rome, Italy); the Apogee SI 400 IR radiometers (Logan, UT, USA) allowed to detect canopy temperature. Longwave infrared radiation was measured by means of a PIR pyrgeometer (Eppley Laboratory, Newport, RI, USA). An ultrasonic anemometer (ATMOS 22, METER Group, Pullman, WA, USA) was used to detect wind speed and direction near the wall. The measurements were taken every 60 s, averaged every 15 min and stored in the data loggers.

2.2 Modelling of the heat fluxes for bare and covered wall

The heat fluxes occurring at the external surface of the BW and the CW were identified, described by mathematical relationships and evaluated by using the gathered experimental data.

The energy balance (B) for both the bare and the covered surfaces was written by including the different heat transfer mechanisms (Fig. 2). Positive values of B mean energy input in the surface.

For the external surface of BW, B was evaluated as:

$$B_{bw} = E_1 - E_2 + \varepsilon_{ws}(R_{sky} + R_g) - R_{e,bw} + CV_{bw,ea}, \qquad (1)$$

where: E terms refer to solar radiation, R terms refer to longwave infrared (LWIR) radiation and CV to convection.

The term E_1 (Wm^{-2}) is the solar radiation on a vertical surface. E_2 (Wm^{-2}) was calculated as a function of the solar reflectivity coefficient of the wall surface (ρ^{sol}_{ws})

$$E_2 = \rho^{sol}_{ws} E_1. \qquad (2)$$

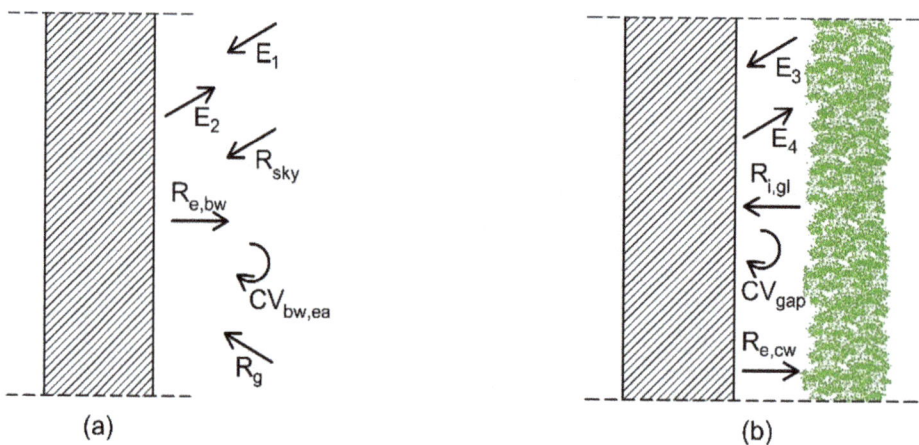

Figure 2: Heat fluxes at the external surface of (a) The BW; and (b) The CW.

The LWIR terms were calculated by:

$$R_{sky} = \sigma F_{sky} T_{sky}^4, \tag{3}$$

$$R_g = \varepsilon_g \sigma F_{g-vs} T_g^4, \tag{4}$$

$$R_{e,bw} = \varepsilon_{ws} \sigma T_{e,bw}^4, \tag{5}$$

where: R_{sky}, R_g and $R_{e,bw}$ (Wm^{-2}) are the LWIR radiative energy coming from the sky and the ground and emitted by the external surface of the BW, respectively; σ is the Stefan-Boltzmann constant equal to $5.67 \cdot 10^{-8}$ W m^{-2} K^{-4}, F_{sky} is the sky view factor between the sky and the building wall and F_{g-vs} is the configuration factor between the ground and the vertical surface, T_{sky}, T_g and $T_{e,bw}$ (K) are temperatures of the sky, of the ground and of the external surface of the BW, respectively; ε_g and ε_{ws} are the emissivity coefficient of the ground and of the BW external surface, respectively.

The convective heat transfer at the external surface of the BW was calculated by:

$$CV_{bw,ea} = h_e(T_{ea} - T_{e,bw}), \tag{6}$$

where h_e (Wm^{-1}K^{-1}) is the external convective coefficient and T_{ea} (K) is the external air temperature.

For the external surface of the CW, B is:

$$B_{cw} = E_3 - E_4 + \varepsilon_{ws} R_{i,gl} - R_{e,cw} + CV_{gap}. \tag{7}$$

The solar radiation terms were calculated by:

$$E_3 = \tau_{gl}^{sol} E_1, \tag{8}$$

$$E_4 = \rho_{ws}^{sol} E_3, \tag{9}$$

where τ_{gl}^{sol} is the solar transmissivity coefficient of the leaves in the GL.

The LWIR radiative energy fluxes were:

$$R_{i,gl} = \varepsilon_{gl} \sigma T_{i,gl}^4 (1 - \rho_{gl}\rho_{ws})^{-1}, \tag{10}$$

$$R_{e,cw} = \varepsilon_{ws} \sigma T_{e,cw}^4 (1 - (\varepsilon_{ws}\rho_{gl})(1 - \rho_{gl}\rho_{ws})^{-1}), \tag{11}$$

where: $R_{i,gl}$ and $R_{e,cw}$ (Wm^{-2}) are the LWIR radiative energy emitted by the inner side of the GL and by the external surface of the CW, respectively; $T_{i,gl}$ and $T_{e,cw}$ (K) are the temperatures of the inner side of the GL and of the external surface of the CW; ε_{gl} and ρ_{gl} are the leaves emissivity and reflectivity coefficients, respectively, and ρ_{ws} is the reflectivity coefficient of the wall surface.

The convective heat flux was calculated as a function of the convective coefficient h_{gap} (Wm^{-1}K^{-1}) [33]:

$$CV_{gap} = h_{gap}(T_{i,gl} - T_{e,cw}). \tag{12}$$

3 RESULTS AND DISCUSSION

The heat fluxes at the external surface of the BW and the CW were evaluated by using the aforementioned mathematical relationships and the experimental data. Calculations were performed for a period of three clear summer days (26–28 August 2019).

Concerning the BW, Fig. 3(a) shows the energy exchanged by means of the three mechanisms at the external surface. During daytime the solar radiation provides the most significant gain of energy. LWIR radiative and convective exchanges are often negative (energy losses) and these essentially allow to disperse energy.

Similar considerations can be done referring to the energy fluxes at the external surface of the CW (Fig. 3(b)). However, in this case, the heat fluxes, especially the solar one, show lower values than the previous ones.

The surface energy budget for the BW and the CW were shown in Fig. 4. These two curves summarize the different behaviour of the two envelope solutions.

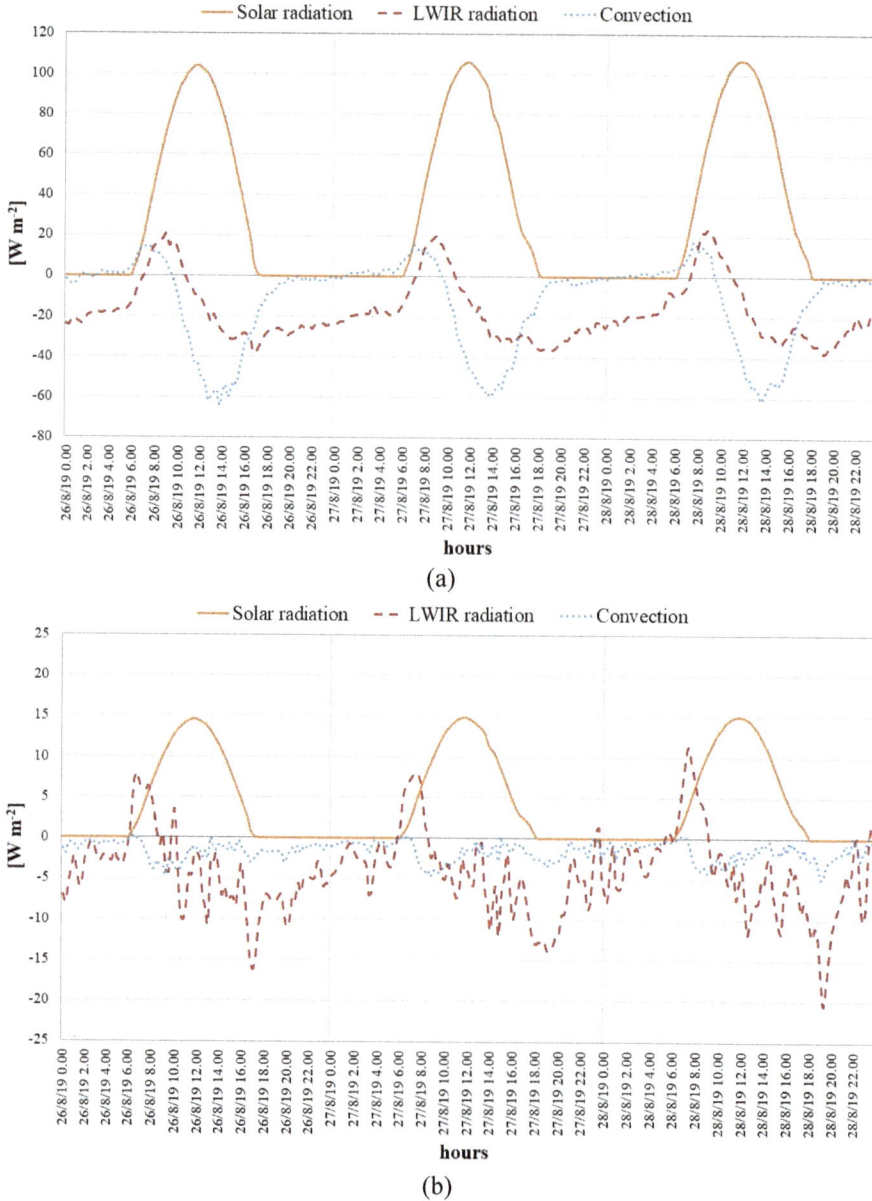

(a)

(b)

Figure 3:   Calculated heat fluxes at the external surface of (a) The BW; and (b) The CW. Solar radiation, LWIR radiation and convection, 26–28 August, 2019.

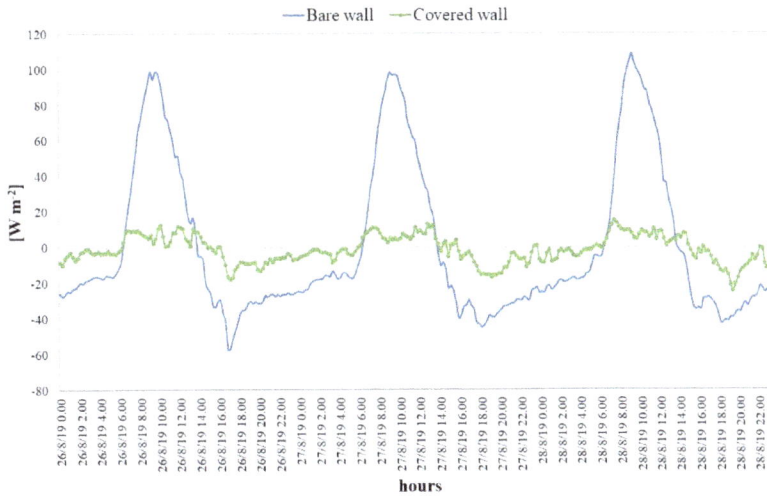

Figure 4:   Energy balance at the external surface of the BW and the CW, 26–28 August 2019.

It is interesting to compare the energy functioning of the BW and the CW, by focusing on the energy budget at their external surfaces (Fig. 5). In Fig. 5 the mean external air temperature was also plotted, to highlight the influence of this climatic parameter on the energy budget. During daytime, the energy gained by the BW was significantly higher than that gained by the CW (Fig. 5(a)). Similarly, the BW lost more energy than the CW at night-time (Fig. 5(b)). If compared to that of the BW, the range of energy variation for the CW was smaller, thus it was more stable. Concerning the energy budget for the all-day periods (Fig. 5(c)), the energy budget for the CW showed similar values for all the three days and it was always negative.

During daytime, the daily mean value of the energy budget for the BW was equal to 1.04 MJm^{-2}, while it was equal to 0.12 MJm^{-2} for the CW. This means that the BW gained 88% more energy than the CW.

At night-time, the BW lost 0.98 MJm^{-2}, while the CW lost 0.25 MJm^{-2}.

Overall, the mean value of the energy budget for the all-day period was equal to 0.06 MJm^{-2} for the BW and to −0.13 MJm^{-2} for the CW.

Concerning the evaluation of the energy performance of the GFs during warm seasons, these findings are significant. It was shown that the GL played a crucial role in reducing the heat fluxes and preventing the heating of the wall behind it. During warm days a BW gains energy, while a CW loses heat.

In the literature, the evaluation of the energy performance of the GFs was mainly based on the analysis of the surface temperature difference between the BW and the CW and on the energy balance at the internal surfaces of the walls. However, the findings of this research are consistent with those of the other studies. In fact, Jim and He [21] and Larsen et al. [36] showed that heat fluxes were significantly reduced in the CW, compared to the BW. Other authors found that the overall heat fluxes were generally negative for the CW and positive for the BW [35], [37], [38] and Chen et al. [37] showed that the CW was cooled especially through radiative heat exchanges. All these authors recognized the relevant advantages of the vertical greenings during summer.

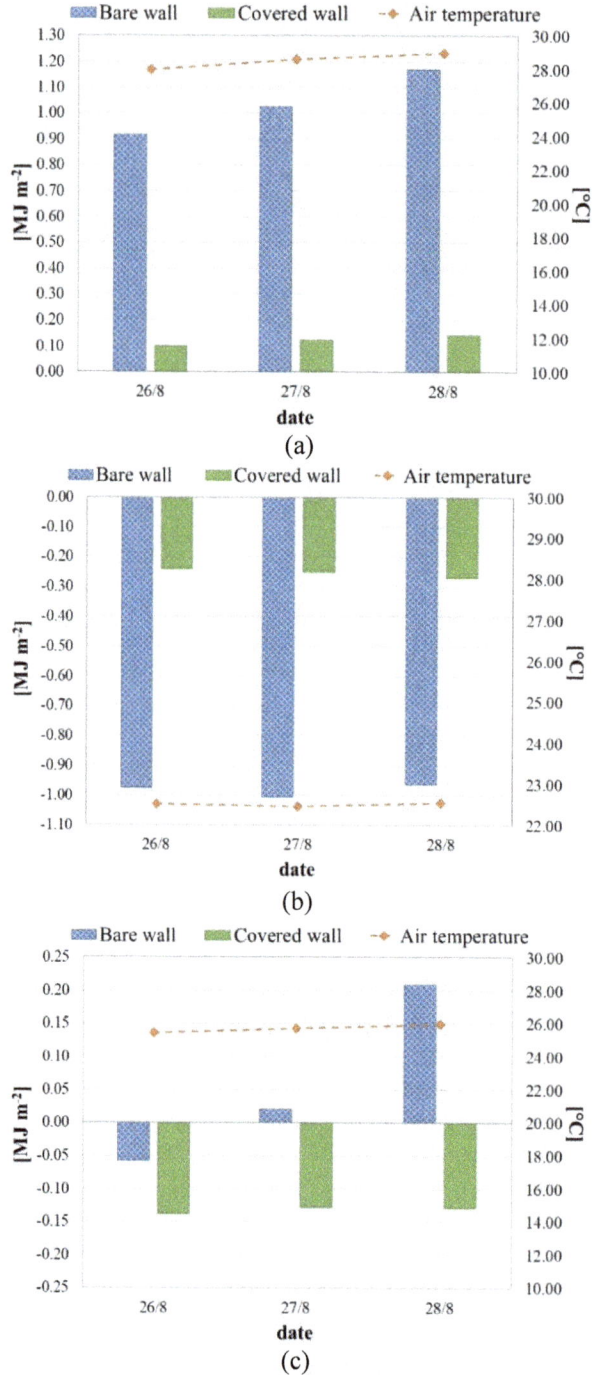

Figure 5:   Energy budget at the external surface of the BW and the CW. (a) Daytime; (b) Night-time; and (c) All day. Mean external air temperature: 26–28 August, 2019.

Nomenclature			
B	Balance of energy (Wm^{-2})	ρ^{sol}	Solar reflectivity coefficient
BW	Bare wall	σ	Stefan-Boltzmann constant
CV	Convective heat transfer (Wm^{-2})		($5.67 \cdot 10^{-8}$ W m^{-2} K^{-4})
CW	Covered wall	τ^{sol}	Solar transmissivity coefficient
E	Solar radiative heat transfer (Wm^{-2})	Subscripts	
F	Configuration factor	bw	bare wall
GF	Green façade	cw	covered wall
GL	Green layer	e	external
h	Convective coefficient (Wm^{-1}K^{-1})	ea	external air
LWIR	Longwave infrared	e,bw	external surface of bare wall
R	LWIR radiative heat transfer (Wm^{-2})	e,cw	external surface of covered wall
T	Temperature (K)	g	ground
UGI	Urban green infrastructure	gl	green layer
VGS	Vertical green system	g-vs	from ground to vertical surface
ε	Infrared emissivity coefficient	i,gl	inner side of green layer
ρ	Infrared reflectivity coefficient	ws	wall surface

4 CONCLUSIONS

Green façades can significantly improve the building energy performance thanks to the application of a layer of vegetation. The presence of plants has a positive influence on microclimatic conditions and on human comfort. In order to produce the best possible benefits and to allow their optimum design, energy models of GFs are needed.

This paper analysed the thermal behaviour of a wall in a GF system and highlighted the advantages compared to a bare wall, in summer clear sky days. Radiative and convective heat transfer were modelled. The evapotranspiration benefits were implicitly considered too. Data coming from an experimental GF, realized at the University of Bari, were recorded and used to calculate the effective heat fluxes at the external surface of the two walls. The solar radiative energy was found to be the most relevant for both walls, but significantly reduced in the covered wall. Convective and LWIR radiative exchanges mainly allowed to disperse energy. The surface energy budget at the outer side of the two building walls were evaluated and compared, taking into consideration the change during the day periods. This study demonstrated that the heat fluxes in a GF system were significantly reduced and that the covered wall was thermally more stable. The energy budget at the surface of the covered wall was negative, this means that it lost energy and that the wall heating was avoided. This energy behaviour has direct consequences on the energy saving for cooling in warm periods. Results obtained by this research can be a useful contribution to the modelling of the energy functioning of GFs and to the knowledge of the real advantages provided by them.

REFERENCES

[1]   Bartesaghi Koc, C., C.P. Osmond, C.P. & Peters, A., Evaluating the cooling effects of green infrastructure: A systematic review of methods, indicators and data sources. *Solar Energy*, **166**, pp. 486–508, 2018. DOI: 10.1016/J.SOLENER.2018.03.008.

[2]   Badiu, D.L., Nita, A., Iojă, C.I. & Niţă, M.R., Disentangling the connections: A network analysis of approaches to urban green infrastructure. *Urban Forestry and Urban Greening*, **41**, pp. 211–220, 2019. DOI: 10.1016/j.ufug.2019.04.013.

[3] Mekala, G.D. & Hatton MacDonald, D., Lost in transactions: Analysing the institutional arrangements underpinning urban green infrastructure. *Ecological Economics*, **147**, pp. 399–409, 2018. DOI: 10.1016/J.ECOLECON.2018.01.028.

[4] Besir, A.B. & Cuce, E., Green roofs and façades: A comprehensive review. *Renewable and Sustainable Energy Reviews*, **82**, pp. 915–939, 2018. DOI: 10.1016/j.rser.2017.09.106.

[5] Raji, B., Tenpierik, M.J. & Van Den Dobbelsteen, A., The impact of greening systems on building energy performance: A literature review. *Renewable and Sustainable Energy Reviews*, **45**, pp. 610–623, 2015. DOI: 10.1016/j.rser.2015.02.011.

[6] Zhao, X., Zuo, J., Wu, G. & Huang, C., A bibliometric review of green building research 2000–2016. *Architectural Science Review*, **62**(1), pp. 74–88, 2019. DOI: 10.1080/00038628.2018.1485548.

[7] Radić, M., Brković Dodig, M. & Auer, T., Green façades and living walls – A review establishing the classification of construction types and mapping the benefits. *Sustainability*, **11**, pp. 4579, 2019. DOI: 10.3390/su11174579.

[8] Medl, A., Stangl, R. & Florineth, F., Vertical greening systems – A review on recent technologies and research advancement. *Building and Environment*, **125**, pp. 227–239, 2017. DOI: 10.1016/j.buildenv.2017.08.054.

[9] Vox, G., Maneta, A. & Schettini, E., Evaluation of the radiometric properties of roofing materials for livestock buildings and their effect on the surface temperature. *Biosystems Engineering*, **144**, pp. 26–37, 2016. DOI: 10.1016/j.biosystemseng.2016.01.016.

[10] Convertino, F., Di Turi, S. & Stefanizzi, P., The color in the vernacular bioclimatic architecture in Mediterranean region. *Energy Procedia*, **126**, pp. 211–218, 2017. DOI: 10.1016/j.egypro.2017.08.142.

[11] Daemei, A.B., Azmoodeh, M., Zamani, Z. & Khotbehsara, E.M., Experimental and simulation studies on the thermal behavior of vertical greenery system for temperature mitigation in urban spaces. *Journal of Building Engineering*, **20**, pp. 277–284, 2018. DOI: 10.1016/j.jobe.2018.07.024.

[12] Ghazalli, A.J., Brack, C., Bai, X. & Said, I., Physical and non-physical benefits of vertical greenery systems: A review. *Journal of Urban Technology*, **26**, pp. 53–78, 2019. DOI: 10.1080/10630732.2019.1637694.

[13] Ling, T.Y. & Chiang, Y.C., Well-being, health and urban coherence-advancing vertical greening approach toward resilience: A design practice consideration. *Journal of Cleaner Production*, **182**, pp. 187–197, 2018. DOI: 10.1016/j.jclepro.2017.12.207.

[14] Blanco, I., Schettini, E., Scarascia Mugnozza, G., Campiotti, C.A., Giagnacovo, G. & Vox, G., Vegetation as a passive system for enhancing building climate control. *Acta Horticulturae*, **1170**, pp. 555–561, 2017. DOI: 10.17660/ActaHortic.2017.1170.69.

[15] Schettini, E., Vox, G., Blanco, I., Campiotti, C.A. & Scarascia Mugnozza, G., Green walls for building microclimate control. *Acta Horticulturae*, **1215**, pp. 73–76, 2018. DOI: 10.17660/actahortic.2018.1215.13.

[16] Hunter, A.M., Williams, N.S.G., Rayner, J.P., Aye, L., Hes, D. & Livesley, S.J., Quantifying the thermal performance of green façades: A critical review. *Ecological Engineering*, **63**, pp. 102–113, 2014. DOI: 10.1016/j.ecoleng.2013.12.021.

[17] Kontoleon, K.J. & Eumorfopoulou, E.A., The effect of the orientation and proportion of a plant-covered wall layer on the thermal performance of a building zone. *Building and Environment*, **45**, pp. 1287–1303, 2010. DOI: 10.1016/J.BUILDENV.2009.11.013.

[18] Widiastuti, R., Caesarendra, W., Prianto, E. & Budi, W.S., Study on the leaves densities as parameter for effectiveness of energy transfer on the green façade. *Buildings*, **8**, pp. 138, 2018. DOI: 10.3390/buildings8100138.

[19] Ip, K., Lam, M. & Miller, A., Shading performance of a vertical deciduous climbing plant canopy. *Building and Environment*, **45**, pp. 81–88, 2010. DOI: 10.1016/j.buildenv.2009.05.003.

[20] Charoenkit, S. & Yiemwattana, S., Living walls and their contribution to improved thermal comfort and carbon emission reduction: A review. *Building and Environment*, **105**, pp. 82–94, 2016. DOI: 10.1016/j.buildenv.2016.05.031.

[21] Jim, C.Y. & He, H., Estimating heat flux transmission of vertical greenery ecosystem. *Ecological Engineering*, **37**, pp. 1112–1122, 2011. DOI: 10.1016/j.ecoleng.2011.02.005.

[22] Xing, Q., Hao, X., Lin, Y., Tan, H. & Yang, K., Experimental investigation on the thermal performance of a vertical greening system with green roof in wet and cold climates during winter. *Energy and Buildings*, **183**, pp. 105–117, 2019. DOI: 10.1016/j.enbuild.2018.10.038.

[23] Olivieri, F., Grifoni, R.C., Redondas, D., Sánchez-Reséndiz, J.A. & Tascini, S., An experimental method to quantitatively analyse the effect of thermal insulation thickness on the summer performance of a vertical green wall. *Energy and Buildings*, **150**, pp. 132–148, 2017. DOI: 10.1016/j.enbuild.2017.05.068.

[24] Koyama, T., Yoshinaga, M., Maeda, K. & Yamauchi, A., Transpiration cooling effect of climber greenwall with an air gap on indoor thermal environment. *Ecological Engineering*, **83**, pp. 343–353, 2015. DOI: 10.1016/j.ecoleng.2015.06.015.

[25] Hoelscher, M.T., Nehls, T., Jänicke, B. & Wessolek, G., Quantifying cooling effects of façade greening: Shading, transpiration and insulation. *Energy and Buildings*, **114**, pp. 283–290, 2016. DOI: 10.1016/J.ENBUILD.2015.06.047.

[26] Blanco, I., Schettini, E. & Vox, G., Predictive model of surface temperature difference between green façades and uncovered wall in Mediterranean climatic area. *Applied Thermal Engineering*, **163**, pp. 114406, 2019. DOI: 10.1016/j.applthermaleng.2019.114406.

[27] Peng, L.L.H., Jiang, Z., Yang, X., He, Y., Xu, T. & Chen, S.S., Cooling effects of block-scale façade greening and their relationship with urban form. *Building and Environment*, **169**, pp. 106552, 2020. DOI: 10.1016/j.buildenv.2019.106552.

[28] Yang, F., Yuan, F., Qian, F., Zhuang, Z. & Yao, J., Summertime thermal and energy performance of a double-skin green façade: A case study in Shanghai. *Sustainable Cities and Society*, **39**, pp. 43–51, 2018. DOI: 10.1016/j.scs.2018.01.049.

[29] Pérez, G., Rincón, L., Vila, A., González, J.M. & Cabeza, L.F., Behaviour of green façades in Mediterranean Continental climate. *Energy Conversion and Management*, **52**, pp. 1861–1867, 2011. DOI: 10.1016/j.enconman.2010.11.008.

[30] Perini, K., Ottelé, M., Fraaij, A.L.A., Haas, E.M. & Raiteri, R., Vertical greening systems and the effect on air flow and temperature on the building envelope. *Building and Environment*, **46**, pp. 2287–2294, 2011. DOI: 10.1016/j.buildenv.2011.05.009.

[31] Sternberg, T., Viles, H. & Cathersides, A., Evaluating the role of ivy (Hedera helix) in moderating wall surface microclimates and contributing to the bioprotection of historic buildings. *Building and Environment*, **46**(2), pp. 293–297, 2011. DOI: 10.1016/j.buildenv.2010.07.017.

[32] Convertino, F., Vox, G. & Schettini, E., Heat transfer mechanisms in vertical green systems and energy balance equations. *International Journal of Design & Nature and Ecodynamics*, **14**, pp. 7–18, 2019. DOI: 10.2495/DNE-V14-N1-7-18.

[33] Convertino, F., Vox, G. & Schettini, E., Convective heat transfer in green façade system. *Biosystems Engineering*, **188**, pp. 67–81, 2019. DOI: 10.1016/j.biosystemseng.2019.10.006.

[34] He, Y., Yu, H., Ozaki, A., Dong, N. & Zheng, S., An investigation on the thermal and energy performance of living wall system in Shanghai area. *Energy and Buildings*, **140**, pp. 324–335, 2017. DOI: 10.1016/j.enbuild.2016.12.083.

[35] Zhang, L. et al., Thermal behavior of a vertical green façade and its impact on the indoor and outdoor thermal environment. *Energy and Buildings*, **204**, pp. 109502, 2019. DOI: 10.1016/j.enbuild.2019.109502.

[36] Larsen, S.F., Filippín, C.S. & Lesino, G., Thermal simulation of a double skin façade with plants. *Energy Procedia*, **57**, pp. 1763–1772, 2014. DOI: 10.1016/j.egypro.2014.10.165.

[37] Chen, Q., Li, B. & Liu, X., An experimental evaluation of the living wall system in hot and humid climate. *Energy and Buildings*, **61**, pp. 298–307, 2013. DOI: 10.1016/j.enbuild.2013.02.030.

[38] Olivieri, F., Olivieri, L. & Neila, J., Experimental study of the thermal-energy performance of an insulated vegetal façade under summer conditions in a continental mediterranean climate. *Building and Environment*, **77**, pp. 61–76, 2014. DOI: 10.1016/j.buildenv.2014.03.019.

SECTION 3
SUSTAINABLE LAND USE

LIFE CYCLE ASSESSMENT OF LEAFY VEGETABLE CONSUMPTION IN URBAN TAIPEI, TAIWAN

MING-HWI YAO[1], HUU-SHENG LUR[2] & CHUN-HSIANG HUANG[2]
[1]Agriculture Research Institute, Council of Agriculture, Taiwan
[2]National Taiwan University, Taiwan

ABSTRACT

In recent years, due to economic development and urbanization, a stable supply of food resources has become a major factor in maintaining normal urban metabolism. However, urban expansion entails a reduction in the cultivable area and a reduced agricultural workforce. Thus, urban areas rely on rural areas for food supply. In this research, Taipei (Taiwan) was selected for investigating changes in the sources of leafy vegetables and the effect of such changes on the environment. The "life cycle assessment (LCA)" method was used to evaluate the effect of different origin-to-destination distances on the environment. The results demonstrated that increased transport distances not only increased energy consumption but also increased the effect on the environment, specifically with respect to the carbon footprint, eutrophication, and acidification. Furthermore, different cultivation methods, including protected, open-field, organic, and conventional cultivation, were compared to determine their environmental effects. The results indicated that the conventional cultivation method increased the carbon footprint by 15 times and energy consumption by 22 times compared with the organic cultivation method. Chemical fertilizer use in conventional cultivation is the main driver of this difference. Moreover, the protected cultivation method exhibited a 10–20 times higher impact on the environment than open-field cultivation for all the indicators in the life cycle assessment, of which its acidification index score was the most seriously problematic. Therefore, urban agriculture and organic cultivation could be promoted to make the best use of the limited land supply in the city and increase the self-sufficiency rate of leafy vegetables cultivation in urban areas and reduce the environmental impact caused by transportation and chemical fertilizer use.
Keywords: urbanization, urban metabolism, life cycle assessment, organic cultivation.

1 INTRODUCTION

As market economies develop, cities provide numerous employment opportunities, and hence, rural populations migrate to cities to make a living; this gradually increases the degree of urbanization. Urbanization across the world began to increase rapidly in the 20th century. In 1900, 1950, and 2000, the degree of global urbanization was 16.4%, 29.1%, and 46.8%, respectively. In 2007, the global urban population exceeded the rural population. Furthermore, the degree of global urbanization reached 54.8% in 2017. It is estimated that by 2050, 68% of the world's population will live in urban areas, and few countries are expected to have larger rural populations than urban populations [1].

Urban metabolism is mainly used to analyze models of matter and energy flow in urban metabolic processes. Wolman [2] suggested that cities are like ecosystems and described how matter and energy flow into the system, similar to how organisms in the ecosystem consume resources, such as sunlight and food, and produce products and waste. When the metabolic system is unable to obtain the resources needed for its internal operation, it must obtain resources from the external environment to support the normal operation of the metabolic system, or it will cause urban metabolic disorders. For example, when a city grows too fast, its managers are unable to maintain control over its various aspects, and thus resource depletion, environmental pollution, ecological damage, and other problems become increasingly prominent. The fundamental reason for these problems is urban metabolic disorder [2], [3].

WIT Transactions on Ecology and the Environment, Vol 243, © 2020 WIT Press
www.witpress.com, ISSN 1743-3541 (on-line)
doi:10.2495/UA200071

Crop cultivation methods can be classified as open-field cultivation and protected culture. In open-field cultivation, crops are grown under natural conditions, allowing large-scale planting and low-cost production. However, many problems must be overcome in open-field cultivation; tasks to overcome problems include ensuring crop quality, minimizing the harmful effects of pests and diseases, decreasing the susceptibility of crops to weather events or climate shifts, and overcoming the inconvenient regulation of the production period. Protected cultivation can overcome the disadvantages of open-field cultivation. In protected cultivation, fresh weight, dry weight, leaf number, and chlorophyll content of leafy vegetables can be increased, thus enhancing yield [4], [5]. However, the production cost of protected cultivation is much higher than that of open-field cultivation; such costs include capital used to construct facilities, the replacement cost of consumables such as plastic cloth, and the energy cost of lamps inside greenhouses. Although the cost of protected cultivation is high, it is a crucial aspect of the food supply chain because it can hedge against food stock disruptions caused by natural disasters and thus provides a stable food supply.

The management methods of crop cultivation can be divided into conventional and organic cultivation methods. The aim of organic agriculture is to augment ecological processes that foster plant nutrition while protecting soil and water resources from damage [6]. Compared with conventional cultivation, organic agriculture is thought to have a less negative impact on the environment because it does not involve the use of chemical fertilizers, synthetic herbicides, or pesticides. Moreover, organic cultivation can enhance the quality of soil, air, and water and also positively affect biodiversity [7]. However, although organic cultivation has a low impact on the environment, its crop yield is usually lower than that of conventional cultivation. The yields of organic cultivation are on average 80% those of conventional cultivation. Therefore, the land required to produce the same amount of food in organic farming systems is usually more than that required in conventional farming [8], [9].

In 2018, the degree of urbanization in Taiwan was 79.5%, and the Taipei metropolitan area had the highest urban population density. Therefore, as the population has grown, the area available for leafy vegetables cultivation in Taipei has gradually decreased, but the demand for leafy crops has gradually increased. Thus, the development of urban agriculture may be necessary and could help in the cultivation of leafy vegetables with a short growth cycle, which is not only cheap but also increased quantities of produce to meet the needs of the people [10].

The impact of food on climate change is divided into two aspects: food production and transportation. The transportation makes a small part of the greenhouse gas emissions created by food; 83% of overall emissions of CO_2 are in production phases [11]. However, "Food miles" is a course indicator of the Greenhouse Gas (GHG) contributions of food transportation. The term "food miles" refers to the distance food travels from farm to consumer. With an increase in the distance that food travels, energy consumption is increased as well as carbon emissions [12]. Taipei metropolitan is the most urbanized area in Taiwan, and food is mostly imported from outside due to insufficient leafy vegetables cultivation area. Therefore, this study also explores the differences in environmental impact under different transportation distances. In addition, to minimize the impact on the environment, the concept of "local food" has been proposed for application in sustainable agriculture and alternative food systems [13]. Local food is considered one means of reducing food mileage.

In summary, this study selected the Taipei metropolitan area of Taiwan, which has the highest degree of urbanization of any city in Taiwan, as the research target and used statistics from government units and the Taipei Fruits & Vegetables Wholesale Market to calculate changes in leafy vegetables production and sources in the Taipei metropolitan area. Furthermore, the method of life cycle assessment (LCA) can be used to assess the integrated environmental impact of the supply chain and of different fields by analyzing a product's (commodity/services) entire life cycle. In an environmental assessment, the impact categories are the main issues discussed; the carbon footprint, acidification, eutrophication, and energy consumption are the main environmental impact indicators, of which eutrophication is mainly caused by fertilizer application during the food production. Then, the impact on the environment is assessed according to space and time; finally, data and findings can inform decision-makers [14]. Therefore, the field survey data were analyzed using the LCA method to assess the different environmental impacts of leafy vegetables in Taiwan produced under different methods of crop cultivation and management and involving different distances of delivery to the Taipei Fruits and Vegetables Wholesale Market, and these data can be used as a basis to solve environmental problem in the future.

2 RESEARCH METHOD

2.1 Analysis of statistical data integration

In the study, we focused our analysis on Taipei, the most urbanized area in Taiwan. According to government unit statistics, the changes in the yield, consumption, and cultivation areas in the Taipei metropolitan area were calculated for the past 30 years. Furthermore, statistical data were used to calculate the space dedicated to leafy vegetables in each area, and the geographic information system (GIS) map was drawn using ArcMap 10.1 to analyze the distribution of the cultivation of leafy vegetables in Taiwan. Additionally, using the statistical data of Taipei Agricultural Products Marketing Corporation, the distribution ratio of leafy vegetables in the Taipei metropolitan area was calculated.

2.1.1 Life cycle assessment
In the study, SimaPro (v 8.4.0.0) was used for LCA of the environmental impact of different cultivation systems, management methods, and transportation distances. The main impact factors assessed were the carbon footprint of greenhouse gas emissions, eutrophication and acidification of water resources and soil caused by irrigation and fertilizers, and nonrenewable energy consumption. In addition, the IMPACT 2002+ V2.12 method was used in SimaPro to estimate the impact index (%).

2.1.2 Life cycle assessment of leafy vegetables from open-field and protected cultivation
The LCA of different cultivation systems starts from the 2 aspects of crop cultivation and facilities. Because the life of facilities is usually 10 to 20 years, an independent inventory must be conducted. The structural components of the growing system primarily consist of steel brackets and plastic sheeting, so our analysis focused on these two items. Steel brackets are used for fixing and a plastic sheet is used for protection in these facilities, and the crops produced are mainly leafy vegetables. The amount of fertilizer used during planting and the energy consumption of the production process were estimated, including energy used for agricultural machinery in land preparation, trenching, fertilizing, and spraying tasks. The energy consumption of agricultural machinery is generally diesel fuel, and the energy unit is expressed in the megajoule (MJ).

2.1.3 Life cycle assessment of leafy vegetables from organic and conventional cultivation
The biggest difference between organic and conventional cultivation is the application of different fertilizers; the remaining processes are subtly different. The amount of fertilizer used during planting and energy consumption under two different management methods were evaluated. Fertilizer application and the use of agricultural machinery and irrigation pumps in the cultivation process generally use diesel fuel for energy, and the energy unit is expressed

in MJ. Additionally, in this study, the results of the environmental impacts of different fertilizer sources were evaluated using the parameters built into SimaPro.

2.1.4 Life cycle assessment of transportation distance

In this study, we focused on the Taipei metropolitan area and took stock of the environmental impact and cost of transporting leafy vegetables to the Taipei Fruits & Vegetables Wholesale Market from different origins. It was assumed that the leafy crops are transported by 2.4-ton trucks using diesel as the energy source. The distance between the main production areas of each county and the Taipei Fruits & Vegetables Wholesale Market was measured using Google Maps and used as the transportation distance for LCA.

2.1.5 Assessment of environmental impact indicators during crop cultivation

In the crop cultivation process, the environmental impact at different stages is not the same. In this study, the four factors that cause an environmental impact are seedling raising, fertilizer use, land preparation (oil consumption by traction machines, cultivators, etc.), and power facilities (electricity consumption by irrigation equipment, etc.) which are used to assess differences in indicators of the carbon footprint, eutrophication, acidification, and energy consumption at each cultivation stage.

3 RESULTS

3.1 Urbanization rate and leafy vegetables supply chain of the Taipei metropolitan area

3.1.1 Leafy vegetables supply chain of the Taipei metropolitan area

Changes in the yield, consumption, and areas of leafy crop cultivation in the Taipei metropolitan area over the past 30 years, determined through analysis of statistical data, are shown in Fig. 1. In 2018, the total area and output of leafy vegetables in Taiwan were 30,948 hectares and 890,000 metric tons, respectively. This study selected 3 regions and analyzed their areas of leafy vegetables cultivation by using the GIS (Fig. 2), and the yields of the three cultivation areas accounted for 35.9%, 11.7%, and 9.2%, whereas the remaining areas accounted for 43.2% (Fig. 3(a)). Furthermore, using the statistical data from the Taipei Fruit & Vegetables Wholesale Market, the proportion of leafy vegetables transported from three areas to the Taipei metropolitan area was calculated (Fig. 3(b)).

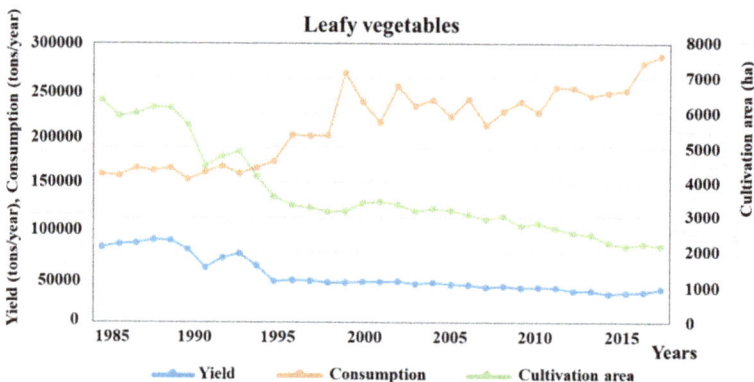

Figure 1: Changes in the yield, consumption, and cultivation area of leafy crops in the Taipei metropolitan area over the past 30 years.

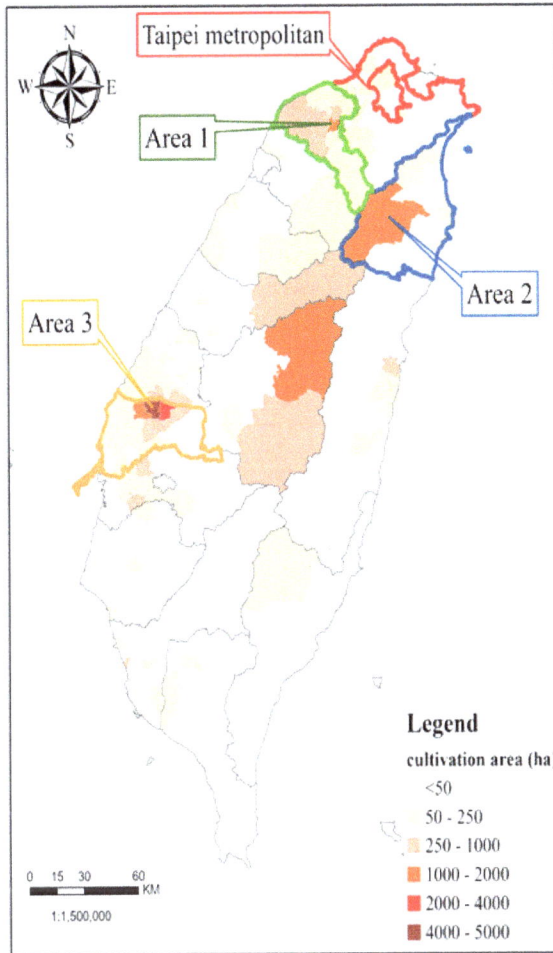

Figure 2: Cultivation areas of leafy vegetables in Taiwan in 2018.

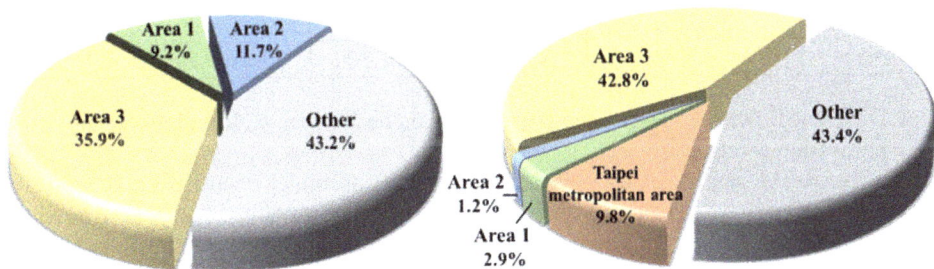

Figure 3:   (a) The proportion of leafy vegetables production in Taiwan in 2018; (b) The proportion of leafy vegetables transported from three areas to the Taipei metropolitan area in 2018.

3.1.2 Environmental impact of different transportation distances

In this study, the Taipei Fruit & Vegetables Wholesale Market was the destination, and three areas of leafy vegetables cultivation were selected for assessment; these areas are referred to as Area 1, Area 2, and Area 3, and the distances of these areas from the Taipei Fruit & Vegetables Wholesale Market are 39.9, 105.5, and 214 km, respectively. The total volume transported from the 3 areas in 2018 was 5.64E+06, 2.31E+06, and 8.46E+07 kg, and the typical transportation vehicle was a 2.4-ton truck. After conversion, the fuel consumption per kilogram of vegetables transportation in the 3 areas was 0.4, 1.1, and 2.3 mL/kg. Thus, the freight mileage (ton/km) of the environmental impact of leafy vegetables transported from three areas to the Taipei Fruit & Vegetables Wholesale Market are shown in Table 1.

Table 1:　Environmental impact of different transportation distances.

Indicators	Unit	Area 1	Area 2	Area 3
Carbon footprint	kg CO_{2eq}	4.32E-03	1.14E-02	2.32E-02
Acidification	kg SO_{2eq}	2.87E-05	7.58E-05	1.54E-04
Eutrophication	kg $PO_{4\ p\text{-lim}}$	3.83E-08	1.01E-07	2.05E-07
Energy	MJ	6.67E-02	1.76E-01	3.58E-01

3.2　Comparison and contrast of cultivation systems

3.2.1　The proportion of environmental impacts at each stage of crop cultivation

In this study, LCA was used to evaluate the rate of carbon footprint, eutrophication, acidification, and energy consumption caused by various aspects of cultivation, including seedling raising, fertilizer use, land preparation, and power facilities, to understand the environmental impact of each cultivation process. The differences among seedling raising, fertilizer use, land preparation, and powering facilities in terms of environmental impact were shown in Table 2.

Table 2:　The proportion of environmental impacts of each part at leafy vegetables cultivation.

Indicators	Unit	Seedling raising	Fertilizer use	Land preparation	Power facilities
Carbon footprint	kg CO_{2eq}	8.53E-03	6.65E-02	7.51E-03	3.78E-02
Acidification	kg SO_{2eq}	3.31E-05	1.30E-04	4.99E-05	1.47E-04
Eutrophication	kg $PO_{4\ p\text{-lim}}$	2.28E-06	1.15E-05	8.03E-08	1.01E-05
Energy	MJ	1.23E-01	1.50E+00	1.08E-01	5.47E-01

3.2.2　Differences between leafy vegetables produced under open-field and protected cultivation practices

The environmental impact of leafy vegetables production under open-field and protected cultivation practices was evaluated on the basis of 1 hectare of cultivation area. The yield, water consumption, and energy consumption of agricultural machinery are shown in Table 3. In addition, in this study, the facility aspect of protected cultivation was calculated independently. In assessing facilities, we used the following values: 31 metric tons of polyvinyl chloride (PVC) plastic sheet and 200 metric tons of steel are required per hectare. The environmental impact of facilities per hectare is shown in Table 4. However, the service life of steel and PVC plastic sheet is approximately 10 years and 5 years, respectively, under normal conditions. Therefore, when evaluating the environmental impact index of steel and plastic sheeting in the LCA, the index must be divided by the service life of steel and PVC plastic sheeting. Furthermore, the environmental impact index of facilities was then combined with the environmental impact index of crop cultivation. The differences between open-field and protected cultivation in terms of environmental impact from the production of leafy vegetables per hectare are shown in Table 5; the environmental impact indicators of protected cultivation are approximately 10–20 times higher than those of open-field cultivation.

Table 3: Field survey data from open-field and protected cultivation.

Data category	Unit	Open-field	Protected
Yields	kg/ha	23,200	21,650
Water consumption	m^3/ha	2,060	1607
Energy consumption	MJ/ha	6,283	10,660

Table 4: Environmental impact of facilities.

Indicators	Unit	PVC plastic sheet	Steel
Carbon footprint	kg CO_{2eq}	2.75E+00	4.71E+00
Acidification	kg SO_{2eq}	1.39E-02	2.86E-02
Eutrophication	kg $PO_{4\ p\text{-}lim}$	4.04E-04	8.07E-04
Energy	MJ	6.81E+01	5.46E+01

Table 5: Environmental impact of open-field and protected cultivation.

Indicators	Unit	Open-field	Protected
Carbon footprint	kg CO_{2eq}	1.04E-01	1.13E+00
Acidification	kg SO_{2eq}	3.27E-04	6.06E-03
Eutrophication	kg $PO_{4\ p\text{-}lim}$	2.16E-05	1.91E-04
Energy	MJ	2.16E+00	2.08E+01

3.2.3 Differences between leafy vegetables produced through organic or conventional cultivation

The environmental impacts of leafy vegetables produced through organic or conventional cultivation were evaluated on the basis of 1 hectare of cultivation area. The yield, water consumption, fertilizer application, and energy consumption of agricultural machinery are shown in Table 6. The differences between organic and conventional cultivation in terms of the environmental impact of the production of leafy vegetables per hectare are shown in Table 7, and the results show that the environmental impact indicators of conventional cultivation are approximately 10–20 times higher than those of organic cultivation. In addition, this study also evaluated the environmental impact of pig manure; cow manure; compost from organic fertilizer; and ammonium sulfate, calcium superphosphate, and potassium chloride from chemical fertilizer; these were all measured per kilogram in SimaPro, and the results are shown in Table 8. In addition, in the LCA calculation software, organic fertilizers are identified as being hand-made by their own farms, and are produced by fermentation in the natural environment without being processed by machines. Therefore, they presume that organic fertilizers have no impact on the environment, as the environmental impact indicators are displayed "N/A"

Table 6: Field survey data on organic and conventional cultivation.

Data category	Unit	Organic	Conventional
Yields	kg/ha	22,500	23,200
Water consumption	m^3/ha	247	2,060
Energy consumption	MJ/ha	941.2	6,283
Ammonium Sulphate			1,235
Calcium ammonium nitrate	kg/ha	N/A	415
Potassium Chloride			415
#43 Instant Water Soluble Fertilizer			825

Table 7: Environmental impact of organic and conventional cultivation.

Indicators	Unit	Organic	Conventional
Carbon footprint	kg CO_{2eq}	6.71E-03	1.04E-01
Acidification	kg SO_{2eq}	2.97E-05	3.27E-04
Eutrophication	kg $PO_{4\,p\text{-lim}}$	1.45E-06	2.16E-05
Energy	MJ	9.69E-02	2.16E+00

Table 8: Data on the environmental impact of organic and chemical fertilizer in SimaPro.

Indicators	Unit	Chemical Fertilizer (kg)		
		Ammonium sulfate	Calcium superphosphate	Potassium chloride
Carbon footprint	kg CO_{2eq}	5.80E-01	2.23E-01	2.07E-01
Acidification	kg SO_{2eq}	1.02E-03	1.14E-03	3.34E-04
Eutrophication	kg $PO_{4\,p\text{-lim}}$	1.30E-05	5.56E-06	4.39E-06
Energy	MJ	1.58E+01	5.94E+00	3.63E+00
		Organic Fertilizer (kg)		
Indicators	Unit	Pig manure	Cow manure	Compost
Carbon footprint	kg CO_{2eq}			
Acidification	kg SO_{2eq}	N/A	N/A	N/A
Eutrophication	kg $PO_{4\,p\text{-lim}}$			
Energy	MJ			

4 DISCUSSION

In Taiwan, limited data are available on the environmental impact of leafy vegetables cultivation and transportation evaluated using LCA. Therefore, this study will establish LCA methods for this part and compare them to explore the differences.

4.1 Influence of food miles on environmental impact

Four basic modes of transporting large quantities of goods are water, railway, road, and aviation. The different modes of transportation result in differences in energy consumption and carbon footprint [11]. Moreover, in addition to transportation methods, transportation distance results in different environmental impacts. Consistent with this study's results (Table 1), another study indicated that when using road transportation methods, the carbon footprint emissions increase as transport distance increases [15]. As mentioned earlier, the greater the energy consumption of the transportation mode and the greater the transportation distance are, the more severe the environmental impact will be. To reduce the negative effects of crop transportation, the concept of urban agriculture has been proposed. The production of food through urban agriculture can reduce the amount of crop transported from a place of origin, thereby reducing food miles and thus reducing greenhouse gas emissions caused by transportation [16].

4.2 Discussion of the environmental impact of various methods of crop cultivation and management

Scholars cultivated lettuce under open-field cultivation and protected cultivation practices, using three types of facilities, namely a plastic greenhouse, plastic mulch, and plastic mulch combined with fleece. The scholars evaluated the environmental impact of different cultivation methods, and the results indicated that protected cultivation has a greater impact on the environment than open-field cultivation, and the results are consistent with the present study (Table 5). In addition, as more facility materials are used, the impact on the environment becomes greater [17]. If the facility aspect of protected cultivation is not considered in the LCA, the difference between open-field cultivation and protected cultivation in terms of environmental impact is negligible. The main cause of the environmental impact of protected cultivation is the facility. Therefore, renewable materials or materials with a long service life should be used to reduce the use of consumables and thereby reduce the impact of facilities on the environment [18].

Scholars cultivated lettuce under organic and conventional cultivation and evaluated the differences in the impact on the environment. Indicators of environmental impact were more serious under conventional cultivation than under organic cultivation [19]. The results are consistent with the present study (Table 7). The biggest difference between organic and conventional cultivation is chemical fertilizer use during cultivation. Therefore, the environmental impacts of chemical and organic fertilizers were evaluated using the parameters built into SimaPro, and the results demonstrated that organic fertilizers in the LCA system do not negatively impact the environment (Table 8). Thus, chemical fertilizers are largely responsible for the impact of conventional cultivation on the environment. However, the previous study has pointed out that nitrogen content is the most critical points influencing environmental impact, but the nitrogen emission model often is built on assumptions from conventional agriculture leading to even greater deviances for organic systems between the amount of N calculated by emission models and the actual amount of N available for emissions. Therefore, in the future, more representative background data on organic cultivation should be established in the LCA to provide a more accurate assessment of organic agricultural cultivation management [10].

4.3 Local production for local consumption

In recent years, the concept of Local Production, Local Consumption (LPLC) has become popular. Importing vegetables from remote areas involves large energy consumption due to transportation. Therefore, by using urban space to grow crops to avoid importing from remote areas, energy consumption can be reduced [20]. This research determined that the area for cultivation of leafy vegetables in the Taipei metropolitan area has gradually decreased but the total demand for leafy vegetables by the people has gradually increased (Fig. 2). Thus, to meet local consumption through local production, developing urban agriculture is a good means of increasing yields of leafy vegetables while reducing energy consumption and carbon footprint. However, local production combined with conventional cultivation has a greater impact on the environment than production in remote areas combined with organic cultivation (data not shown). Therefore, it suggested that utilizes local production should combine with organic cultivation to reach the goal of reducing the environmental impacts.

4.4 Application and positive effects of urban agriculture

Urban agriculture is a method of implementing agricultural cultivation in or around the city and can overcome food availability problems associated with a large urban population. However, the development of urban agriculture may be limited by other forms of urban development, especially in high-density cities; thus, the concept of rooftop farming has been proposed to solve the problem of insufficient land. Rooftop farming provides several benefits, including food supply, carbon sequestration, stormwater retention, energy savings, and mitigation of the urban heat island effect [16]. Moreover, to produce a large amount of food on limited land, vertical cultivation is a good solution that improves land-use efficiency, increases yield per unit area, and increases crop yield [21]. Through the development of urban agriculture, it is possible to increase crop yields in the urban area to meet the needs of the large population of the city and also to reduce the number of crops transported from origin sites, thereby reducing carbon emissions.

5 CONCLUSIONS

With increased urbanization and urban expansion, the areas of leafy vegetables cultivation in the Taipei metropolitan area have gradually decreased but the demand for leafy vegetables has gradually increased. Therefore, to meet people's demand for leafy vegetables in the Taipei metropolitan area, they are imported from other areas to the Taipei metropolitan area. However, this importation from distant production areas results in a severe environmental impact.

In addition, in the process of producing leafy vegetables in various areas, different methods of cultivation and management are used, and each has its advantages and disadvantages. According to this study's results, compared with open-field cultivation and organic cultivation, protected cultivation and conventional cultivation have a greater impact on the environment. However, protected cultivation can shorten the crop cultivation cycle to increase the multiple cropping index and increase yield. Moreover, facilities can protect crops from the impact of natural disasters and stabilize the crop supply.

In summary, if urban agriculture can be developed in Taipei metropolitan area, especially in the form of vertical farming, the limited land supply in the city can be used as much as possible and increase the yield per unit of land. In addition, organic cultivation can also be promoted to replace chemical fertilizers with homemade compost. Through the aforementioned methods, it may be possible to solve the problem of insufficient production areas to increase the self-sufficiency rate of leafy vegetables production in the Taipei metropolitan area and reduce the number of crops transported from distant production areas. Therefore, the demand of Taipei metropolitan area for leafy vegetables can be met while reducing the environmental impact caused by the transportation of leafy vegetables and chemical fertilizer use.

ACKNOWLEDGEMENTS

We thank Professor Fi-John Chang (National Taiwan University) for help with this research. This research (107-2621-M-002-004-MY3) was funded by the Ministry of Science and Technology.

REFERENCES

[1]   Ritchie, H. & Roser, M., *Urbanization. Our World in Data*, 2018.
[2]   Wolman, A., The metabolism of cities. *Scientific American*, **213**(3), pp. 178–193, 1965.
[3]   Zhang, Y., Urban metabolism: A review of research methodologies. *Environmental Pollution*, **178**, pp. 463–473, 2013.
[4]   Travieso, L.L., Leon, A.P., Logegaray, V.R. & Frezza, E.D., Loose leaf lettuce quality grown in two production systems. *European Journal of Scientific Research*, **12**(30), pp. 55–66, 2016.
[5]   Carey, E.E., Jett, L., Lamont, W.J., Nennich, T.T., Orzolek, M.D. & Williams, K.A., Horticultural crop production in high tunnels in the United States: A snapshot. *HortTechnology*, **19**(1), pp. 37–43, 2009.
[6]   Pimentel, D., Hepperly, P., Hanson, J., Douds, D. & Seidel, R., Environmental, energetic, and economic comparisons of organic and conventional farming systems. *BioScience*, **55**(7), pp. 573–582, 2005.
[7]   Lorenz, K. & Lal, R., Environmental impact of organic agriculture. *Advances in Agronomy*, Elsevier Inc., pp. 99–152, 2016.
[8]   De Ponti, T., Rijk, B. & Van Ittersum, M.K., The crop yield gap between organic and conventional agriculture. *Agricultural Systems*, **108**, pp. 1–9, 2012.
[9]   Meier, M.S., Stoessel, F., Jungbluth, N., Juraske, R., Schader, C. & Stolze, M., Environmental impacts of organic and conventional agricultural products – Are the differences captured by life cycle assessment. *Journal of Environmental Management*, **149**, pp. 193–208, 2015.
[10] De Bon, H., Parrot, L. & Moustier, P., Sustainable urban agriculture in developing countries. A review. *Agronomy for Sustainable Development*, **30**(1), pp. 21–32, 2010.
[11] Weber, C.L. & Matthews, H.S., *Food-Miles and the Relative Climate Impacts of Food Choices in the United States*, 2008.
[12] Saunders, C. & Barber, A., Carbon footprints, life cycle analysis, food miles: Global trade trends and market issues. *Political Science*, **60**(1), pp. 73–88, 2008.
[13] Lang, T. & Heasman, M., *Food Wars: The Global Battle for Mouths, Minds and Markets*, Routledge, 2015.
[14] Chang, D., Lee, C.K.M. & Chen, C.H., Review of life cycle assessment towards sustainable product development. *Journal of Cleaner Production*, **83**, pp. 48–60, 2014.
[15] Wakeland, W., Cholette, S. & Venkat, K., Food transportation issues and reducing carbon footprint. *Green Technologies in Food Production and Processing*, pp. 211–236, 2012.
[16] Whittinghill, L.J. & Rowe, D.B., The role of green roof technology in urban agriculture. *Renewable Agriculture and Food Systems*, **27**(4), pp. 314–322, 2012.
[17] Romero-Gámez, M., Audsley, E. & Suárez-Rey, E.M., Life cycle assessment of cultivating lettuce and escarole in Spain. *Journal of Cleaner Production*, **73**, pp. 193–203, 2014.

[18] Anton, A., Montero, J.I., Munoz, P. & Castells, F., LCA and tomato production in Mediterranean greenhouses. *International Journal of Agricultural Resources, Governance and Ecology*, **4**(2), pp. 102–112, 2005.
[19] Foteinis, S. & Chatzisymeon, E., Life cycle assessment of organic versus conventional agriculture. A case study of lettuce cultivation in Greece. *Journal of Cleaner Production*, **112**, pp. 2462–2471, 2016.
[20] Hara, Y., Tsuchiya, K., Matsuda, H., Yamamoto, Y. & Sampei, Y., Quantitative assessment of the Japanese "local production for local consumption" movement: A case study of growth of vegetables in the Osaka city region. *Sustainability Science*, **8**(4), pp. 515–527, 2013.
[21] Eigenbrod, C. & Gruda, N., Urban vegetable for food security in cities. A review. *Agronomy for Sustainable Development*, **35**(2), pp. 483–498, 2015.

ASSESSING THE PERFORMANCE OF URBAN GREEN INFRASTRUCTURE: THE CASE STUDY OF BENICALAP DISTRICT, VALENCIA, SPAIN

CARLA M. TUDORIE[1], MARÍA VALLÉS-PLANELLS[2], ERIC GIELEN[3] & FRANCISCO GALIANA[2]
[1]Universitat Politècnica de València (UPV), Spain
[2]School of Agricultural Engineering and Environment, Department of Rural and Agrifood Engineering, Universitat Politècnica de València, Spain
[3]School of Civil Engineering, Department of Urbanism, Universitat Politècnica de València, Spain

ABSTRACT
During the last few years, the number of cities which are making efforts in improving urban greenery as a measure to strengthen urban resilience and citizen's wellbeing is increasing. The assessment of the implementation of urban green infrastructure actions, as any other type of intervention, involves the use of indicators to measure the effects of these actions. These indicators are understood as parameters that allow assessing the impact and the temporary monitoring of the effects of human actions carried out on the territory. Drawing on the literature of ecosystem/landscape services and green infrastructure, this study proposes a set of indicators which are able to analyse the performance of urban green infrastructure in terms of structure and functionality. A selection of the proposed indicators connected to structural properties, regulation and cultural landscape services are tested at the street level in Valencia City, Spain. Indicators are applied in two pilot actions which are being developed in Benicalap District, Valencia, in the context of a broader project which aims to give evidence about the benefits of nature-based solutions. This research contributes to a better understanding of how indicators can be used as an effective tool to assess the landscape services provided by urban green infrastructure. This information can be useful to communicate the benefits of urban green infrastructure and to make decisions about the development of sustainable strategies based on urban greenery.
Keywords: nature-based solutions, sustainability, urban greenery, landscape services, Valencia.

1 INTRODUCTION
Green areas deliver a wide range of landscape services (LS) to face city's challenges and reach resilience and sustainability. Nowadays, there is an increasing concern of the relevance of giving evidences of the benefits offered by green infrastructure [1]–[4], in order to prove that an increase of urban green spaces would provide a proportionally larger number of ecological, economic and social benefits and services. Urban green infrastructure (UGI) includes green, partly green and blue spaces (e.g. wetlands). It also includes green spaces which can be located close to the city or within the city itself, like agricultural land [2].

During the last years, European Commission has shown a growing concern in the role of green infrastructure (GI) and many reports have been developed to help find solutions to urban challenges through the implementation of GI. The Commission has started to pay attention to the need of efficient tools to map GI components, which are thought to support the future landscape services' assessment. Some examples of physical mapping methods are available (e.g. Guidos Toolbox, Conefor software package, Linkage Mapper, Quickscan) and have been deployed to city case studies from the local to pan-European scales [5]. Despite the benefits associated with UGI, implementation is far from straightforward or to be incorporated in applied urban planning and management [6], [7].

Sustainable strategies could be successfully designed and implemented through guiding current and hypothetical results of UGI improvements with the aid of a set of green performance indicators. This research brings knowledge of the impacts of nature-based

WIT Transactions on Ecology and the Environment, Vol 243, © 2020 WIT Press
www.witpress.com, ISSN 1743-3541 (on-line)
doi:10.2495/UA200081

solutions (NBS) related to three urban challenges, which were proposed by EKLIPSE experts [3] to guide the design, development, implementation and assessment of NBS pilot projects in urban and in the climate mitigation context [4].

The main goal of this work is to propose a set of green performance indicators and show how to practically assess the landscape services provided by urban green infrastructure. These examples of urban green infrastructure indicators, adapted to landscape services classification, are applied for the evaluation of environmental and cultural benefits provided by pilot actions in open spaces in Benicalap District, Valencia. Key performance indicators (KPIs) are applied to pre-greening and post-greening stages of Grow Green NBS pilot projects. In particular, this work focuses on the monitoring of two of these actions: a sustainable small forest and a green-blue corridor.

2 METHODOLOGY

2.1 Proposal of urban green indicators

Considering the relevance of green indicators and urban green infrastructure for urban planning [7], [8], a set of the most mentioned UGI indicators is proposed (Table 1) to be applied at urban level to assess UGI performance.

The proposal of indicators is structured in two parts following the structure–function–value chain model proposed by Termorshuizen and Opdam [9]. First, the structure indicators are related to spatially explicit landscape characteristics, which may be connected to more than one landscape service. For instance, large tree diameter generally involves a large tree canopy that promotes stormwater interception, reduction of local temperature and habitat for nesting birds and small mammals. Secondly, the function and value indicators are proposed for assessing the principal regulation and cultural landscape services delivered by UGI. The function indicators refer to the potential or capacity to deliver a service (e.g. runoff coefficient in relation to precipitation quantities or recreation potential) and the value indicators are connected to the benefit, economic or non-economic, for people (e.g. economic benefit of reduction of storm water to be treated in public sewerage system). The functional and value indicators are arranged according to the LS and their subcategories following previous work on indicators, for example, Pakzad and Osmond [8] and Valls-Donderis et al. [10]. LS are classified according to the framework established by Vallés-Planells et al. [11], which was based in the Common International Classification of Ecosystem Services (CICES) [12].

2.2 Location of nature-based solutions in city of Valencia:
   Grow Green projects study area

Benicalap District has 2,216,000 m^2 of surface of which 367,038 m^2 are urban green infrastructure. The resident population reaches 46,699 inhabitants, around 8 m^2 of green spaces per inhabitant [13]. The shortest distance between green spaces is 228 m, which facilitates 62.5% of inhabitants have access to urban green infrastructure in the very nearby (at 50 m from their house) and 100% of Benicalap population further (at 300 m, 500 m distance).

Grow Green project (GG) embeds five NBS-actions [4] in Ciutat Fallera neighbourhood, whose objective is to make citizens aware of the need of sustainability in the cities. Ciutat Fallera and Benicalap are neighbourhoods within Benicalap District (Fig. 1) northwest of the city. These actions seek to restore the ancestral connections and ensure the transition between

Table 1: Collection of landscape services and green indicators (KPIs) at urban level.

GI indicators linked to structural characterization			
	Theme	UGI indicators	
Structure	Spatial structure	Green area/inhabitant, proximity to green spaces, total area of green space, accessibility of urban green, structural and functional connectivity, impervious surface, facilities	
	Urban vegetation structure	Density of trees by street section, percent of street trees of top most abundant genus, family and order, tree diameter classes, tree crown coverage	

GI indicators linked to regulation and maintenance landscape services			
	Class	Group	Subgroup	UGI indicators
Function and value	Flow regulation	Water flow regulation	Runoff regulation	Runoff coefficient in relation to precipitation quantities, economic benefit of reduction of storm water to be treated in public sewerage system, flood peak reduction
	Regulation of physical environment	Atmospheric regulation	Climate regulation	Temperature reduction in urban areas, reduced building energy use for heating and cooling
			Climate change mitigation	Total amount/yearly carbon sequestration and stored in vegetation
			Air quality regulation	Annual amount of pollutants (O_3, NO_2, SO_2, PM_{10}, $PM_{2.5}$) captured by vegetation
			Noise regulation	Noise level attenuation
	Regulation of biotic environment	Lifecycle maintenance and habitat protection	Biodiversity maintenance and pest and disease control	Vegetation and wildlife diversity, habitat heterogeneity, species suitability

GI indicators linked to cultural landscape services			
	Class	Group	Subgroup	UGI indicators
Function and value	Health	Physical health		Walkability, increase of physical outdoor activity
		Mental health		Reduced depression and anxiety, attention restoration, recovery from stress
	Enjoyment			Recreation potential, green space quality, green space attractiveness, green space visitation
	Self-fulfilment	Didactic resources		Outdoor educational activities
	Social fulfilment	Social relationship		Social interactions, community activities, social cohesion
		Place identity		Neighborhood attachment

the rural and the urban ecosystems. In Valencia, the urban orchard, Huerta de Valencia, is a unique agroecosystem, declared World Heritage by the UN, and a fundamental component of UGI and maybe a sustainable strategy to tackle climate mitigation. The target actions of

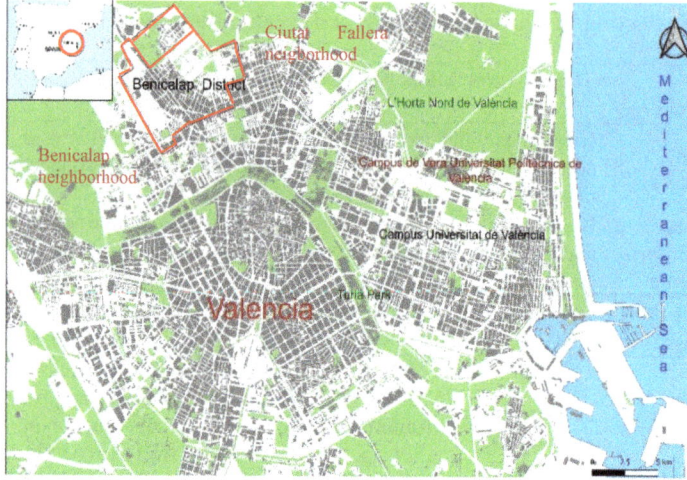

Figure 1: The urban green infrastructure and the location of Benicalap District in Valencia, Spain.

this study are: a green-blue corridor (GG action 3) and a sustainable small forest (GG action 2).

Two sites of Benicalap District are selected for practical assessments (Fig. 2). A street connected to a small square as a part of a corridor and a poor in vegetation-vacant lot of Ciutat Fallera were selected for the location of NBS pilot projects. Four alternatives were assessed to choose the most suitable location for the green-blue corridor (Fig. 2). The sustainable small forest was decided to be created in the neighbourhood of Benicalap park and two farmhouses (Alquería del Moro and Alquería de La Torre).

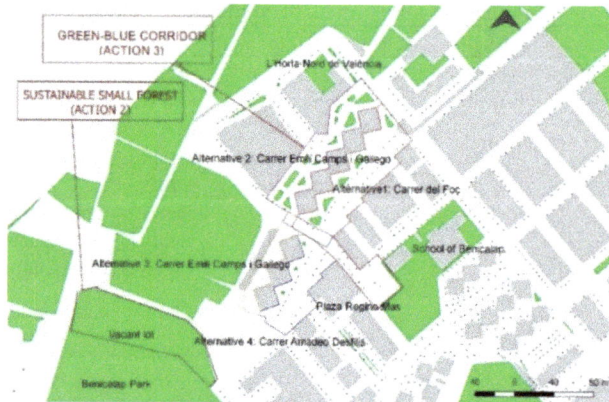

Figure 2: The current state of UGI of Ciutat Fallera, Benicalap, the locations of NBS pilot projects: Sustainable small forest (Action 2) and green-blue corridor (Action 3); Alternative 1: Carrer del Foc and Plaza Regino Mas, Alternative 2: Carrer Emili Camps Gallego (north) and Plaza Regino Mas, Alternative 3: Carrer Emili Camps Gallego (south) and Plaza Regino Mas, Alternative 4: Carrer Amadeo Desfilis and Plaza Regino Mas.

2.3 UGI classification: Selection of landscape services provided by UGI to assess

The most relevant green indicators found in literature are applied to street elements of urban green infrastructure. For this purpose, UGI map is developed with the aid of GIS and fieldwork (Fig. 2). Table 2 shows the UGI street elements included in Nature-based solution pilot projects which were proposed in the context of Grow Green H2020 European project.

Table 2:   UGI elements found in pilot projects in Benicalap neighbourhood according to European Union [2], Clément [14] and Valencia City Council [15].

Vegetation cover	UGI	Description
1. Ruderal vegetation	Abandoned, ruderal areas	Recently abandoned areas, construction sites, etc. with spontaneously occurring pioneer or ruderal vegetation.
2. Green street, avenues and boulevards	Tree alleys and street trees, hedges	Trees having or not having tree pits, planted along roads and paths either solitary or in rows, trees surrounding hedges along roads or paths.
	Green verge	Stripes of green, e.g. flowers, along a built or natural element.
	Street green	Non-tree, mostly shrubby or grassy road verges or areas between the opposite roadways.
	Green island	Sustainable green roundabout, which can contain flowers, shrubs and trees.

2.4 Analysis blocks and UGI indicators used for pilot projects assessment

The sequence of urban green infrastructure indicators is shown according to their utility to achieve study's objectives and is structured in three blocks (Table 3):

A. Selection of the most suitable alternative for the green corridor. Four vegetation plan alternatives (Fig. 2) are assessed as possible locations of a green-blue corridor according to UGI biodiversity indicators.
B. Assessment of the impact of the green-blue corridor and the sustainable small forest on regulation and maintenance LS: biodiversity. This block includes current evaluations (pre-greening stage) and future (post-greening stage) of indicators connected to plant community structure and biodiversity to demonstrate the effectiveness of UGI implementation and green indicators' validity.
C. Assessment of the impact of the green-blue corridor and the sustainable small forest on cultural LS. Social interactions and physical activity level are assessed before the implementation of NBS through cultural UGI indicators.

2.5 Urban vegetation structure indicators description

The density of trees by street section is a value that relates the number of trees per meter of transverse section. Some minimum criterion of tree density per section are used, so the optimum density for big size trees should be 0.2 [16]. Longitudinal transects along pilot projects were done in June 2019. Considering the length of project actions, street section

Table 3: Proposed indicators to assess the performance of the green infrastructure of Benicalap District, Valencia, Spain. The division of indicators is according to their utility case.

Subgroup	UGI indicators	Utility
Urban vegetation structure	Density of trees by street section, Percent of street trees of top most abundant genus, family and order, Tree diameter classes, Crown coverage of the tree (large) Index of abundance of street trees in pilot projects	B
Biodiversity maintenance and pest and disease control	Diversity of trees, shrubs and bushes, diversity of vegetation strata, Species suitability	A, B
Increase of physical activity	Percentage of people undertaking forms of physical activity	C
Community cohesion	Percentage of people interacting with each other in an outdoor space	C

measures 50 m. Tree diameter classes indicator is used to classify green infrastructure trees according to their size class distribution. It is desirable for cities to manage and reach standard values of street trees distribution in UGI elements. Street trees are classified in size classes according to tree diameter breast height (DBH) [17].

Crown coverage is the projected surface of the tree canopy cover that was used to calculate habitat diversity. Performance indicators are proposed to obtain a desirable percentage of existing urban tree canopy by using street trees canopy desired values as proposed by Leff [18].

Index of abundance of street trees is used to evaluate street trees species taxonomic order present in the urban ecosystem. Current species richness of street trees of pilot projects is assessed with UGI map. Information about street trees species is found on the website of Statistical Office of Valencia municipality.

2.6 Biodiversity

In order to calculate the biodiversity of trees, shrubs and bushes, Shannon–Wiener Index (H') is used, with the aid of Braun-Blanquet scale [19]. According to Krebs [20], evenness (E) oscillates between 0 (species are not equally distributed) and 1 (there is equal species distribution, complete evenness). Urban green infrastructure diversity of vegetation strata is also estimated with Shannon–Wiener Index based on five different proportions of habitat proposed by Handley [21] to measure the heterogeneity of a habitat: bare ground and turf grass, rough grassland and herbs, shrubs, trees (tree canopy) and built (impervious) environment. Diversity strata index of post-greening moment is an estimation since the real surface covered by new introduced vegetation is uncertain.

Protective tree diversity can be measured with the aid of three indicators: age diversity or age classes considering diameter classes, species diversity and species suitability [18]. Species suitability is understood as their level of adaptation to local climate, fulfil soil, humidity and management's requisites and are not invasive.

2.7 Physical activity and social interactions: Mohawk

Method for observing physical activity and wellbeing, on short Mohawk, proposed by Benton et al. [22] was designed to measure physical activity and social interactions in small urban green space such as: pocket parks, tree-lined streets and green corridors along waterways. In this study, we apply these cultural indicators to assess present situation of pilot projects.

Social interactions were measured with behavioural observations of individuals who are interacting with the people around them. The same person can be coded as engaging in multiple behaviours. Physical health of Ciutat Fallera's inhabitants was assessed with indicators of physical activity level. Physical activity is more than physical exercise [23]. Regarding the adequacy of field observations, data are collected about inhabitants "age and gender and ruled hours" intervals. Pre-greening assessment of neighbourhood wellbeing indicators was carried out during one week on November 2019.

3 RESULTS AND DISCUSSION

3.1 Selection of the most suitable alternative for the green corridor

The analysis of the four green corridor alternatives (Fig. 2) in terms of biodiversity indicates that alternatives 1 and 2 are the best locations according to biodiversity. This information was used together with data on indicators related to other regulation services (runoff regulation, climate regulation and climate change mitigation), cultural services (recreation potential and place identity) and other factors connected to gender issues in order to decide the best alternative. After this process, alternative 1, Carrer del Foc street and Plaza Regino Mas square, was selected as the best location to build green-blue corridor (Table 4).

Table 4:   Results of diversity indices applied to vegetation species and habitat types of proposed alternatives for selecting green-blue corridor emplacement.

	Diversity indices	Alternatives			
		1	2	3	4
Vegetation	Shannon–Wiener Index	1.93	1.98	1.76	1.71
	Maximum diversity	3.78	3.66	3.76	3.76
	Evenness	0.51	0.54	0.47	0.45
Strata	Shannon–Wiener Index	0.99	1.07	0.94	0.98
	Maximum diversity	1.61	1.61	1.61	1.61
	Evenness	0.61	0.66	0.59	0.61
% Habitat cover	Bare ground/turf grass	9.18	12.67	9.93	9.35
	Rough grassland/herb	0.41	0.64	0.24	0.17
	Shrub, bush, creeper	5.10	6.52	3.69	3.26
	Tree	65.85	62.74	67.53	62.64
	Built environment	19.45	17.43	18.61	24.57

3.2 Assessment of the impact of the green-blue corridor and the sustainable small forest on regulation and maintenance LS: biodiversity

3.2.1 Structural indicators
There are 8,899 individuals of 113 species spread all over Benicalap District and 122 individuals of 11 species of street trees in the pilot projects. Within street trees species of

NBS pilots there are 88.52% deciduous trees and 11.48% perennial trees. The area of pilot projects represents 0.57% of Benicalap District, where 11 species of seven families and six orders are concentrated (Table 5).

Table 5: Number of individuals, species, index of abundance (I_{TREES}) and percentage of street trees species at district, street and pilot projects (actions) level.

Order	District	Street	I_{trees}	Species	District %	Street %	Pilot %
Arecales	1689	1	0.00	*Phoenyx dactylifera*	3.78	0.30	0.85
Asparagales	71	1	0.01	*Yucca elephantipes*	2.86	2.38	0.85
Fabales	790	9	0.01	*Ceratonia siliqua*	0.54	12.50	0.85
				Cercis siliquastrum	25.68	1.59	5.93
				Robinia pseudoacacia	4.63	1.47	0.85
Lamiales	1366	48	0.04	*Jacaranda mimosifolia*	27.72	9.58	33.05
				Olea europaea	2.59	23.68	7.63
Malpighiales	52	1	0.02	*Populus sp.*	3.54	1.92	0.85
Sapindales	1825	35	0.02	*Ailanthus altissima*	0.20	100.00	2.54
				Melia azedarach	29.97	7.27	27.12
Urticales	33	23	0.70	*Ulmus pumila*	2.25	72.73	19.49

Density of trees by street section is 0.1, which means that the recommended distances [16] are not always respected (Table 5). Present number of street trees is two times smaller than it should. Recommended indicator has value of 0.2 for 100 m longitudinal transect, which means one tree at every five meters.

Size class distribution is applied specifically to street trees. In both pilot projects, semi-mature trees dominate (68.12%, 50.00%), while senescent trees are very poor represented (1.45%) or are missing (0.00%) (Table 6). The other two age classes are also unbalanced. According to Leff [18], the ideal percentages of street trees distribution are not respected (40% juvenile, 30% semi-mature, 20% mature and 10% senescent trees). In the pilot projects, there are almost double and more than double semi-mature trees than it should be (30%). More juvenile trees are needed in both cases, as they revitalize the urban ecosystem. In the corridor, the percentage of mature trees reaches around 20%. However, in the forest, the existing proportion (29.17%) overpasses the recommendation (20%). Our study results are opposite to McPherson et al. findings of cities of Inland Valley, California, where street inventories were characterized by higher percentage of young trees populations (55%) [24]. The similitude appears for lower than ideal percentages of mature (18%) and old (5%) trees.

Present relative tree canopy cover of the corridor represents 69.20% of total area (7,352 m²), which is qualified as fair [18]. Present forest tree canopy cover is estimated as low (24.40% of total surface (5,356 m²)) (Table 6). Low status sites investigated by Pauleit et al. [25] presented around 3% of tree cover of 0.25 km². Considering Ciutat Fallera's financial situation [13], we could it classify as low status neighbourhood.

3.2.2 Biodiversity indicators

The estimation of the values of post greening vegetation diversity shows a significant rise of these indicators in both actions. In the case of the green-blue corridor, Shannon–Wiener Index increases from 1.15 to 1.93, which means that goes from an initial poor biodiversity level (0.5 < H' < 1.5) to almost a good level (2 < H' < 3) according to Agencia d'Ecologia

Table 6: Community structure indicators of the current situation of street trees within project actions.

Indicators		Green-blue corridor	Sustainable small forest
Street trees density (N. trees/50 m)	Canopy width (m)		
	Small size	8	0
	Medium size	2	0
	Big size	0	5
Age diversity (%)	Juvenile	14.49	20.83
	Semi-mature	68.12	50.00
	Mature	15.94	29.17
	Senescent	1.45	00.00
Relative tree canopy (%)		69.20	24.40

Urbana de Barcelona [16] and evenness raises in 50% (Table 7). Regarding the forest which, at present time, contains just one species (*Ulmus pumila*), post greening stage involves reaching a very good level of biodiversity (3 < H' < 4) and a more equal species distribution (closer to value 1). According to Barker et al. [6], community's accessible green spaces, street trees and hedgerows are very likely to enhance biodiversity at micro scale.

Table 7: Diversity of vegetation species, strata diversity and habitats proportions before and after selecting the pilot projects to construct the green-blue corridor (action 3) and small sustainable forest (action 2).

	Diversity indices	Action 3		Action 2	
		Pre-greening	Post-greening	Pre-greening	Post-greening
Vegetation	Shannon–Wiener Index	1.15	1.93	*	3.07
	Maximum diversity	3.40	3.78	*	4.13
	Evenness	0.34	0.51	0	0.74
Strata	Shannon–Wiener Index	0.97	0.99	0.56	0.69
	Maximum diversity	1.61	1.61	1.61	1.61
	Evenness	0.60	0.61	0.35	0.43
% Habitat cover	Bare ground/turf grass	7.43	9.18	74.60	
	Rough grassland/herb	0.04	0.41	–	80.00
	Shrub, bush, creeper	2.44	5.10	–	
	Tree	57.99	65.85	25.40*	20.00
	Built environment	32.09	19.45	–	0

*At pre-greening stadia there is only a set of decrepit *Ulmus pumilla.*

When comparing the diversity of vegetation strata for pre and post greening stages, the increase in Shannon–Wiener Index (from 0.97 to 0.99) and evenness (from 0.60 to 0.61) is slight in the case of the corridor. Regarding forest, similar values are found (Shannon–Wiener Index rises from 0.56 to 0.69 and evenness from 0.35 to 0.43). This result can be explained by the fact that in the pre-greening stage, only two out of five types of habitats were identified (trees, 25.4% and bare ground/turf grass, 74.6%), while, after the implementation of NBS, three categories of habitats, such as: bare ground and turf grass, rough grassland and herbs and shrubs, bushes and creeper plants are expected to increase around 80%. This result is the estimation of the first year post greening. When trees reach their maturity, crown coverage is expected to be more than 70%. In 2000, all sites of Merseyside, UK, the structural diversity (Shannon–Wiener Index) is 1.09 for a vegetation area which covers 8.31 ha [25].

Among the species of pilot project, five street tree species out of ten are identified as being suitable for the area. According to Leff [18], this means a fair degree of adaptability (50%) to Valencia's climate and soil conditions. These species are: Mediterranean native species (*Cercias siliquastrum, Ceratonia siliqua, Olea europaea*), naturalized species, useful for inhabitants e.g. date palm (*Phoenix dactylifera*) and ornamental species cultivated for decorative purpose e.g. blue jacaranda (*Jacaranda mimosifolia*). All of them are present in the location of the corridor in the pre-greening stage.

3.3 Assessment of the impact of the green-blue corridor and the sustainable small forest on cultural LS

Fig. 3 shows the results of behavioural indicators of social interactions and physical activity measured before the implementations of NBS with Mohawk tool. Results on social interactions reveal that 34.77% of people were talking with each other or having other types of physical contact e.g. holding hands in the street part of corridor, 28.89% in the square part. And 29.93% persons interacted in the forest area. This can be explained by the fact that Carrer del Foc street and Plaza Regino Mas is four times smaller than forest area. Also, corridor's square presents some facilities, such as: bus stop, bars, restaurants, playgrounds and school, which leaded to casual encounters, reflected by the big percentage of neighborhood community who was passing through and having contact in the green-blue corridor. Also, Plaza Regino Mas has a smaller surface than forest. The small surface of forest was used mostly to walk dogs and pass through. According to Barker et al. [6], neighbourhood green spaces are likely to improve social interactions and enhance physical health of inhabitants. Regarding indicators of physical activity level, it can be seen that more than 80% of people were walking in the corridor's arena, while in the forest area were estimated almost 72%. A percentage of 10.51% of vigorous people were crossing the forest area, while 2.65% were walking fast or running in corridor's street and 12.28% in the square part. Forest pilot project is a calmer open space, an isolated one with lower traffic than corridor, which is more dynamic and characterized by a big number of passing cars and noise. The creation of green-blue corridor and sustainable forest are expected to increase both social interactions and level of physical activity.

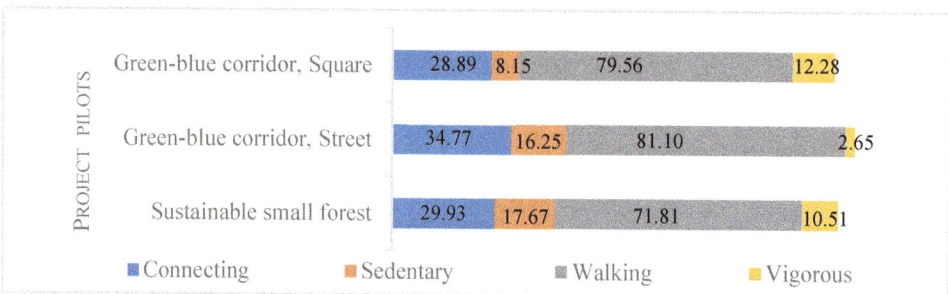

Figure 3:   Values of behavioural KPIs for one week at sustainable small forest action and green-blue corridor action (street and square). Bar chart numbers represent percentages of people who are connecting and undertaking forms of physical activity.

The indicators applied in this work are limited to the pre-greening stage of two small scale pilot projects with poor impact at district scale. Further studies should consider the whole range of UGI elements, including agriculture land which was not part of the analysed case studies.

4 CONCLUSIONS

The article reviews the applicability of green infrastructure indicators at micro scale or street scale and tests the usefulness of a set of these indicators to design and evaluate the effects of NBS implementation and the improvement of green infrastructure. Not only structure indicators are used to assess UGI condition, but also indicators linked to regulation and cultural landscape services, especially to biodiversity and people behaviour.

The assessment of regulation services, in terms of structural indicators, suggests that none of the sites where the pilot projects will be developed meet the recommended density of street trees and tree age diversity. Tree canopy is fair in the green-blue corridor and low in the sustainable small forest for the pre-greening stage. Concerning biodiversity indicators, they allow the prediction of the post-greening stage situation, showing an increase in Shannon–Wiener Index and evenness, which is especially significant for the sustainable small forest. The assessment of the cultural services connected to social interactions and physical activity indicates that the green-blue corridor shows slightly higher values than the sustainable small forest does at the pre-greening stage.

This paper demonstrates that UGI indicators are a helpful method, complementary to cartography, in order to assess design alternatives and select the most sustainable future strategies to be implemented at urban scale. By unifying green infrastructure indicators and methods of measurement from three complementary domains, such as ecology, forestry and social-cultural sciences, this work contributes to a better understanding of UGI as a sustainable tool and to make better informed decisions regarding its implementation in urban planning and management.

The results of this study show how urban green infrastructure elements can be used to integrate urban and agroecosystems. Working in a space of opportunity in its transition between city and orchard, new elements of the urban green infrastructure can be integrated and used to extend the network and solve the interface between agriculture and city, contributing thus to deliver a wide range of ecosystem services' in both rural and urban settings, the purpose of the green infrastructure.

ACKNOWLEDGEMENTS

This work was supported by the European Union's Horizon 2020 research and innovation programme under the project Green Cities for Climate and Water Resilience, Sustainable Economic Growth, Healthy Citizens and Environments with reference 730283. We are grateful to Las Naves, Ayuntamiento de Valencia and Bipolaire for data collaboration.

REFERENCES

[1]   De Ridder, K. et al., An integrated methodology to assess the benefits of urban green space. *Science of the Total Environment*, **334–335**, pp. 489–497, 2004. DOI: 10.1016/j.scitotenv.2004.04.054.

[2]   European Union, Urban GI components inventory milestone 23. Green Surge Project, 2013. www.greensurge.eu. Accessed on: 25 Sep. 2018.

[3]   European Union, Horizon 2020. EKLIPSE Expert Working Group on Nature-based Solutions, An impact evaluation framework to guide the evaluation of nature-based solutions projects. (Draft) 19, 2016. www.eklipse-mechanism.eu/apps/Eklipse_data/ website/EKLIPSE_Report1-NBS_FINAL_Complete-08022017_LowRes_4Web.pdf. Accessed on: 12 Mar. 2020.

[4]   Grow Green, Grow Green, a partnership for greener cities to increase liveability, sustainability and business opportunities. www.Growgreenproject.eu. Accessed on: 4 Mar. 2020.

[5]   European Commission, Strategic green infrastructure and ecosystem restoration. Geospatial methods, data and tools, 2019. www.ec.europa.eu/environment/nature/ ecosystems/index_en.htm. Accessed on: 1 Apr. 2020.

[6]   Barker, A. et al., Understanding green infrastructure at different scales: A signposting guide. University of Manchester: Manchester, 2019. DOI: 10.13140/RG.2.2.26366.95045.

[7]   Tudorie, C.M., Gielen, E., Valles-Planells, M. & Galiana, F., Urban green indicators: A tool to estimate the sustainability of our cities. *Urban Agriculture 2018 1st International Conference on Urban Agriculture and City Sustainability*, New Forest, UK. 2019. DOI: 10.2495/DNE-V0-N0-1-11.

[8]   Pakzad, P. & Osmond, P., Developing a sustainability indicator set for measuring green infrastructure performance. *Procedia: Social and Behavioral Sciences*, **216**, pp. 68–79, 2016. DOI: 10.1016/j.sbspro.2015.12.009.

[9]   Termorshuizen, J.W. & Opdam, E.P., Landscape services as a bridge between landscape ecology and sustainable development. *Landscape Ecology*, **24**, pp. 1037–1052, 2009. DOI: 10.1007/s10980-008-9314-8.

[10]   Valls-Donderis, P., Vallés, M.C. & Galiana, F., Criteria and indicators for sustainable forestry under Mediterranean conditions applicable in Spain at the forest management unit scale. *Forest Systems*, **24**(1), e004, 2015. DOI: 10.5424/fs/2015241-05542.

[11]   Vallés-Planells, M., Galiana, F. & Van Eetvelde, V.A., Classification of landscape services to support local landscape planning. *Ecology and Society*, **19**(1), p. 44, 2014. DOI: 10.5751/ES-06251-190144.

[12]   Haines-Young, R. & Potschin, M., Common International Classification of Ecosystem Services (CICES) V5.1 Guidance on the Application of the Revised Structure, UK, 2017. https://cices.eu/content/uploads/sites/8/2012/07/CICES-V43_Revised-Final_ Report_29012013.pdf. Accessed on: 25 Mar. 2020.

[13]   Ayuntamiento de Valencia, Oficina de estadística, Barrio Ciutat Falera. www.valencia.es/ayuntamiento/webs/estadistica/inf_dtba/2019/Distrito_16_Barrio_ 2.pdf. Accessed on: 16 Mar. 2020.

[14]   Clément, J., *Manifesto del tercer paisaje*, Gustavo Gili, 2007.

[15]   Ayuntamiento de València, Propuesta de plan especial verde para la ciudad de Valencia, Valencia, Ayuntamiento de Valencia, 181, unpublished document, 1992.

[16]   Agencia d'Ecología Urbana de Barcelona, Plan de indicadores de sostenibilidad urbana de Vitoria-Gasteiz. Ayuntamiento de Vitoria-Gasteiz, 2010. www.vitoria-gasteiz.org/wb021/http/contenidosEstaticos/adjuntos/es/89/14/ 38914.pdf. Accessed on: 10 Mar. 2020.

[17]   Gering, L.R. & May, D.M., The relationship of diameter at breast height and crown diameter for four species groups in Hardin County, Tennessee. *Southern Journal of Applied Forestry*, **19**(4), pp. 177–181, 1995. DOI: 10.1093/sjaf/19.4.177.

[18] Leff, M., The Sustainable Urban Forest a Step-by-Step Approach, Davey Institute/ USDA Forest Service USFS Philadelphia Field Station, 2016. www.itreetools.org/ resources/content/Sustainable_Urban_Forest_Guide_14Nov2016.pdf. Accessed on: 5 Mar. 2020.

[19] Braun-Blanquet, J., *Plant Sociology*, McGraw-Hill: New York, p. 539, 1932.

[20] Krebs, C.J., Species diversity measures. *Ecological Methodology*, Harper and Row Inc.: New York, 1989.

[21] Handley, J.F., *Nature in the Urban Environment: City Landscape*, Grove Cresswell, Butterworths: London, pp. 47–59, 1988.

[22] Benton, J.S., Anderson, J., Cotterill, S., Hunter, R.F., Pulis, M. & French, D.P., Method for observing physical activity and wellbeing (MOHAWK): Observation Manual. University of Manchester: Manchester, 2018.

[23] World Health Organization, Global strategy on diet, physical activity and health. www.who.int/dietphysicalactivity/factsheet_adults/en/. Accessed on: 2 Apr. 2020.

[24] McPherson, E.G., van Doorn, N. & de Goede, J., Structure, function and value of street trees in California, USA. *Urban Forestry and Urban Greening*, **17**, pp. 104–115, 2016. DOI: 10.1016/j.ufug.2016.03.013.

[25] Pauleit, S., Roland Ennos, R. & Golding, Y., Modeling the environmental impacts of urban land use and land cover change: A study in Merseyside, UK. *Landscape and Urban Planning*, **71**, pp. 295–310, 2005. DOI: 10.1016/j.landurbplan.2004.03.009.

PROPOSAL OF A SET OF INDICATORS FOR SUSTAINABILITY EVALUATION OF FOOD PRODUCTION IN AN URBAN CONTEXT

MARCIO ALEXANDRE ALBERTI[1], ILEANA BLANCO[2], GIULIANO VOX[1],
LUCIENE PIMENTEL DA SILVA[3] & EVELIA SCHETTINI[1]
[1]Department of Agricultural and Environmental Science DISAAT, University of Bari, Italy
[2]Department of Biological and Environmental Sciences and Technologies DiSTeBA, University of Salento, Italy
[3]Post Graduation Program in Environmental Science, Rio de Janeiro State University, Brazil

ABSTRACT
Urban agriculture (UA) is steadily gaining supporters worldwide, and this is partly because constantly growing urban populations recognize the need to increase food production and promote a healthy diet. UA can also generate extra income, promote socialization leading to valorisation of urban areas. Nevertheless, UA faces several challenges, including limited available space, stressed natural resources, pollution in a context characterized by climate change and new consumption patterns. At the same time, if UA is practiced in the same way as other forms of agriculture, it may have some undesirable environmental impacts on urban areas. The use of indicators can provide decision makers with a tool to evaluate the sustainable insertion of agriculture in cities. This work aims to propose a minimum set of indicators as the basis for an index to assess progress of urban vegetable production towards sustainability. Nineteen urban sustainability indexes, composed of several indicators founded in the literature and already used by several institutions, were analysed. These indexes were broken down into a set of 1579 indicators. Analysis of the relevancy to issues such as food, energy, water, land and pollution, and the selection based on the criteria of being measurable, sensitive to stress, predictable, anticipatory, controllable, integrative, responsive and stable, led to a 26 basic indicators selection. These are proposed as the basis for an innovative UA sustainability index. This will be built based on the Delphi method and is intended to support communities in establishing sustainable and resilient cities.
Keywords: urban agriculture, vegetables, NEXUS, food, energy, water, land use, pollution.

1 INTRODUCTION
Urban agriculture (UA) can be defined as "farming operations taking place in and around the city that beyond food production provides environmental services (soil, water and climate protection; resource efficiency; biodiversity), social services (social inclusion, education, health, leisure, cultural heritage) and supports local economies by a significant direct urban market orientation" [1]. UA initiatives are highly articulated to many of the 2030 UN 17 sustainable development goals (SDGs) [2]. These have been adopted by non-governmental organizations and public administrators as a sustainable and ecologically correct solution to the problems of food security, job and income generation in the metropolis. UA has gained even more support in the context of urban heat island mitigation and climate change, as a direct contribution towards adaptive management. Food production within UA also provides a number of benefits such as the reduction of emissions related to food transport, improvement in the nutritional level of the population, and the use of rainwater. UA can also contribute to the spread of Sustainable Urban Drainage Systems (SuDS) in the context of sustainable drainage as in nature-based solutions, sponge cities, and blue-green cities.

Agriculture is a topic inextricably linked with water, energy and land use. Factors that cannot be approached separately, such as food production, water, energy and land use, are studied jointly in what several world organizations refer to as the "nexus" approach [3]. Nexus approaches are conceptions and initiatives introduced, preferably transversely, to achieve an ecological and circular economy, allowing actions aimed at investing in the conservation of ecosystem services, creating more with less, and accelerating the social and economic improvement of excluded groups by promoting the inclusion of poorer populations [4].

However, as with other forms of agriculture, UA can harm the environment, and naturally requires more water, energy, and space within an already crowded metropolis, where natural resources are often already stressed. UA must be developed while respecting the local capacity to achieve sustainability. "Indicators" are tools that can provide an early warning of problems and allow adequate preventive action [5]. These are widely used in understanding, monitoring and managing complex systems where simplification is required, such as an urbanized area or a natural ecosystem [6]. An ecological indicator is a simplification of the complex ecosystem and interlinkages, which has a wide application in the development of policies, plans and programs related to the environment and ecology via measurement processes, allowing assessment of progress toward sustainable development [7].

Numerous government and private institutions are working on urban sustainability indicators, but few instruments able to measure and assess progress towards the sustainability of food production in cities are available in the literature, and these are often scattered and difficult to access [8]. This work assumes that sustainable UA (target situation) can produce food for all the inhabitants of a city by using available local resources (water, energy, land), without harming the natural flows of resources and without contaminating the environment. Many authors also assume that UA will provide only certain kinds of food for population needs [9], [10]. However, UA may also involve cultivations that adapt well to small spaces and often to alternative cultivation areas, such as roofs, walls, underground galleries or indoor environments in buildings [11]. UA can take advantage of producing foods with high market value and productivity, such as vegetables.

This study aimed to propose a minimum set of indicators as the basis for creating an index to assess the progress of urban vegetable production towards sustainability. This is part of a wider research project aimed at defining an index based on this set of indicators of sustainability for urban vegetable production. The intention is that this index will be formulated in the future using the Delphi method, involving the consultation of specialists in this field. They will be questioned about the suitability of the proposed indicators, and about the relative importance of each one, and this will be taken into account in the different weighting of each indicator in the formulation of the index.

2 MATERIAL AND METHODS

The present work was based on a conceptual framework of three levels of selection, which aimed to identify a set of basic indicators (also called concrete or specific indicators) able to assess progress toward sustainable UA. A group of indexes used by the international community to assess urban sustainability was identified in the scientific and technical literature as the source for proposing the indicators to constitute an index for assessment of the sustainability of these UA systems.

We mainly used a set of indexes gathered in the European Commission report entitled "In-depth Report – Indicators for Sustainable Cities" [12]. The indexes were analysed and broken down into a list of basic indicators. This initial group evaluates various aspects related to urban development such as quality of life, water management, climate change, housing, transportation, risk resilience and the entire food system. The list includes 15 sustainable indexes developed by institutions in Europe, China and the USA, and by supranational institutions, such as the World Bank, the United Nations and Local Governments for

Sustainability – ICLEI (a group of more than 1,750 local and regional governments engaged in sustainable urban development).

Four additional indexes were identified in the literature and these were added to the set of indexes collected by the European Commission: "The City Region Food System" [13] and "The Milan Urban Food Policy Pact" [14], both related to the whole food systems; a conceptual framework from SUSFAN, a team of European experts on sustainable food and nutrition security, which developed a set of concepts, metrics and tools for helping policy and decision makers assess the progress towards goals related to the EU food system [15]; and the SDGs from the UN [2]. The complete list of 19 indexes and a description of the relative features of each one is shown in Table 1.

Table 1: List of urban and food production sustainability indexes.

Sustainability indexes	NI	Indicators' main areas	Subject of assessment
1 – China Urban Sustainability Index (USI 2013) [12]	23	Society, Environment, Economy	Quality of life (Benchmark = New York)
2 – City Blueprint [16]	24	Human wellbeing, Environment, Economic wellbeing	Urban water management (quick scan and baseline assessment)
3 – Blue City Index (BCI*) [17]	25	Water quality, Solid waste treatment, Basic water services, Wastewater treatment, Infrastructure, Robustness, Governance	Urban water management (quick scan and baseline assessment)
4 – European Green Capital Award [12]	12	Green growth and eco-innovation, Sustainable land use, Waste management, Sustainable urban mobility	Improvement of the urban environment and progress towards healthier and sustainable living areas
5 – Green City Index [18]	30	CO_2 emissions, Energy, buildings, Land use, Transport, water and sanitation, Waste management, Air quality and Environmental governance	Environmental performance
6 – Global City Indicators Program (Facility) [19]	115	City services, Quality of life	City governance, City services and quality of life
7 – Global City Indicator – ISO 37120 [20]	128	Education, Energy, Transport, Health, Water management	Smartness and resilience
8 – Indicators for Sustainability [21]	32	Economy, Environment, Social	City progress toward sustainability
9 – Reference Framework for Sustainable Cities (RFSC-v3) [22]	30	Spatial dimension, Governance, Society and culture, Economy, Environment	Project/program of sustainability

Table 1: Continued.

Sustainability indexes	NI	Indicators' main areas	Subject of assessment
10 – STAR (2.0)– (Sustainability Tools for Assessing and Rating Communities) [23]	49	Built environment, Climate and energy, Economy and jobs, Education and arts, Equity, empower, health and safety, Natural systems, innovation and process	Economic, environmental and social aspects
11 – Urban Audit [12]	333	Demography, Social aspects, Economic aspects, Civic involvement, training and education, Environment, travel and transport, Information, society, culture and recreation	Life quality
12 – Urban Ecosystem Europe [24]	25	Health, Consumption, Mobility, energy and climate change, Social equity, Governance	Environmental sustainability and climate change
13 – Urban Metabolism Framework [25]	15	Urban flows, Urban patterns, Urban drivers, Urban quality	The sustainability of a city
14 – Index of urban sustainability performance [26]	16	Society, Environment, Economy	Economic, environmental and socio-cultural progress
15 – Habitat Agenda Indicators [27]	42	Land, Social development and eradication of poverty, Environmental management, Economic development, Governance	Cities without Slum – commitment to the world's poor
16 – Sustainable Development Goals [2]	232	Human factor, Food security, Water, Energy, Infrastructures	Achievement of 2030 Agenda for Sustainable Development objectives
17 – City Region Food System [13]	210	Food security, Economy, Natural resources, Social aspects	Whole food system, from production to waste, food system policy and planning.
18 – MILAN PACT [14]	42	Governance, Sustainable diets and nutrition, Social and economics equity, Food production, Food supply and distribution, Food waste	Whole food system, from production to final consumer
19 – SUSFAN [15]	196	Food, Environment, Economy, Human factor	Whole food system, from production to final consumer

(NI) number of basic indicators used in the INDEX.

The first selection was made by considering the relationship of each basic indicator with at least one component of the Nexus Food-Energy-Water-Land Use-Pollution. This includes environmental, economic and social criteria. We defined this Nexus, inspired by the following: Food-Energy-Water (FEW) [28], Water-Energy-Land-Food (WELF) [29], Ecosystems-Water-Food-Land-Energy (EWFLE) [30] and Energy-Water-Pollution [31] Nexus. Moreover, redundant indicators or those unrelated to UA were removed.

Urban indicators linked to safety, governance, public policy, culture, sport and recreation were not considered in the context of UA and food production. Transport indicators were not considered either, since one benefit of urban agriculture is to produce food locally, thus reducing transport costs ("zero miles food").

A second selection was made according to the recurrence of the indicator (or its variants) in the 19 first selected indexes, so that we selected the basic indicators considered by at least two sustainability indexes.

This second set of potential indicators/indexes underwent a third selection process. This selection required indicators to have certain "individual qualities", i.e. they must be measurable, sensitive to stress, predictable, anticipatory, controllable, integrative, responsive and stable [32], [33].

3 RESULTS AND DISCUSSION

The breakdown process performed on the 19 indexes for assessment of urban sustainability resulted in a set of 1579 basic indicators.

The first selection, concerning relevancy to the "Nexus" and to UA, gave 512 basic indicators. In order to facilitate this study phase, the Nexus categories were split into 9 sub-categories, as illustrated by the first two columns of Table 2. Food was split into food production characteristics, and human factors, including work to produce food and the socio-economic conditions that influence this (education level and income level). The energy category was sub-divided into energy source and CO_2 emission. Water was sub-divided into water sources and effluent issues. Land-use was sub-divided into a "characteristics" group that aggregates space-related indicators, like soil occupation and use (built up area ratio, green space, disposable agriculture area), and another group that addressed residues/waste issues (waste collection and recycling ratio). Finally, pollution was sub-divided into three sub-categories: air, water and soil. Since there were already two special sub-categories to group indicators regarding water (sources and effluents), and no basic indicator met the requirements of considering soil pollution, it was established that pollution would be linked only to indicators that affect air quality (Table 2).

The second selection concerning indicator recurrence in the indexes reduced the number of indicators to 102.

The third and final selection involved all the indicators which possessed all the individual qualities. This resulted in 26 indicators, as shown in Table 2, which also shows the metric units for each indicator.

Knowledge of food production is necessary in relation to the target of self-sufficiency for the local population in terms of the recommended daily intake of 300g of vegetables per person per day [34]. Total local production of vegetables is given in tons/period. The adoption of a food production system that uses natural fertilizers (possibly by recycling the city's organic waste) and natural pest and disease controls would favour UA sustainability. The fraction of local production of vegetables produced without chemicals is given as a ratio of the total.

Table 2: Selected basic indicators for sustainability evaluation of food production in an urban context.

Category	Sub-category	N.	Basic indicator	Unity
Food	Production	1	Total vegetable production	tons/time
		2	Total vegetable production without chemicals	%
	Human factor	3	Human Development Index (HDI)	–
		4	Income commitments on food	%
		5	Unemployment rate	%
		6	Gini Index	
		7	Residential/Individual Internet access	%
		8	Adult illiteracy	%
		9	Obesity incidence on childhood/population	%
Energy	Sources	10	Residential /individual energy consumption	kWh/cap
		11	Energy price	US$/kWh
		12	Energy green source ratio	%
	CO_2 emission	13	Emission CO_2 equivalent per habitant	tCO_2/cap
Water	Source	14	Annual Rainfall	mm
		15	Residential/individual water consumption	L/day/cap
		16	Water access rate	%
		17	Water quality (Incidence of Non-Standard Total Coliform Analysis)	%
		18	Public water supply price	US$/m³
	Effluents	19	Population served by wastewater collection	%
		20	Sewage treatment	%
Land use	Characteristics	21	Land tenure ratio	%
		22	Protected green space	%
		23	Undeveloped land use ratio	%
	Waste	24	Population served by regular solid waste collection	%
		25	Municipal waste recycling	%
Pollution	Air	26	Exceedance of air quality standards in urban areas	day/year

The Human Development Index (HDI) reflects the per capita income level, education and life expectancy of a given population. Although there is a strong correlation between wealth and good performance in sustainability indexes [18], Moran et al. [35] found that countries with higher incomes have greatly improved their HDI at the expense of a larger

Ecological Footprint, while poorer countries have developed sufficiently without harming the environment, thus pointing to an inconsistency between HDI improvement and sustainable development.

The indicators "Income commitments on food" and "Unemployment rate" are basic indicators intrinsically linked to sustainable food production within the city. Data reveal that the poorest classes spend up to 85% of their income on food [36].

The Gini index is proposed as an indicator for the category of food. It is a measurement of income inequality, which ranges from 0 (perfect equality) to 1 (maximum inequality). In many countries that well performed on Gross Domestic Product (GDP), it is actually concentrated in a small sector of the population. The Gini coefficient offers a more balanced view of the progress required for an entire population to reach the target of self-reliance in vegetables established in this study.

Factors influencing the effectiveness and speed of project development are the ratio of access to the web, and illiteracy. Access to the web is essential for the development of smart cities, where the use of technologies based on the Internet of Things (IoT) will help to save energy, making traditional production processes more efficient and cities more resilient to climate change [37]. The web can be used to access new knowledge and technologies, to acquire inputs and to commercialize products.

Illiteracy can contribute to the unsustainability of UA. Nchanji et al. [38] described a case study from Northern Ghana on local government's distribution of pesticides for use on the region's crops. Since most urban farmers were illiterate, they were unable to read the instructions on the packaging labels, and used pesticides inappropriately, thereby contaminating themselves and the environment.

Obesity is an important quality indicator linked to the food factor of the Nexus. Although it does not reflect the lack of access to calories, obesity reflects access to low quality calories, which is often due to the lack of economic power to acquire calories and/or the existence of "food deserts", and UA has the potential to diversify and improve the diets of urban citizens [39]. Borges et al. [40] identified inequalities in the geographical distribution of food retailers in Jundiaí city, Brazil, which commercialize healthy and/or unhealthy foods. Local production via UA could improve the situation of many poor city areas with limited access to adequate food at affordable prices.

Concerning the category of energy, Schnell [41] and Eigenbrod and Gruda [42] have pointed out that factors such as off-season production and the low scale of urban production require higher energy use for food production. The higher demand for energy could be met by using alternative green energy sources, such as solar power, wind, geothermal and biomass. Therefore, it is fundamentally important to use indicators that quantify the total consumption of households, energy pricing and alternative energy use in the total energy matrix of a community.

With regard to emissions, the use of food production technologies is highly dependent on energy in cities at high latitudes or in desert areas. Greenhouses with heating or cooling systems can have a negative impact on the environment and thus overshadow the benefits of zero miles food [43]. Indicators that measure the amount of CO_2 released into the environment per inhabitant can reflect the extent to which societies and their related activities are (or are not) dependent on non-renewable energy [44]

Annual rainfall is an indicator related to the most sustainable source of water for agriculture. This has become a critical factor in a changing climate, since a change in rainfall patterns has been observed in many cities, with more intense and longer spells of rain [45]. Lupia et al. [46] have demonstrated that some of the UA developed in Rome could be supplied by rainwater collected from roofs. The use of indicators related to water quality is just as important as those related to water availability. Water contaminated by chemical or biological agents used in the irrigation of fruits and vegetables (mainly those that are consumed raw) are important vectors of disease among the population [47].

Use of the public water supply can be the safest source, sometimes the only/easiest one, for the irrigation of vegetables in urban areas. However, it can be a criticality in cities that are reaching their maximum supply capacity, and many cities already have problems with network leakages accounting for up to 40% of the total. There is also the problem involved with the management of waste and effluents. The inclusion of indicators regarding household access to a piped public water supply, inhabitants' consumption and water prices are justified in a future sustainability Index.

Regarding space, land ownership plays a key role in UA. Most of the ground surface in cities is covered or sealed by buildings, roads or pavements [48], so that new agricultural production systems are required in order to adapt to the built areas. In addition, urban land tends to be "super" valued. The need to pay for tenure is also a key issue for the sustainability of UA projects. In addition, land does not necessarily consist of fertile soil, and this is another production factor for UA. The ratio of protected green space can be interpreted as an important sustainability indicator, since these green areas offer important environmental services that indirectly favour UA [49]. Another possibly good indicator is the amount of undeveloped land that could be cultivated, showing the feasibility of using these spaces for UA, although many urban areas may have problems such as heavy metal contamination of the soil.

However, contaminated sites may not be totally compromised, since many works point to bioremediation of these soils with the addition of organic matter and phosphates, or even substitution of the more superficial soil layers for the production of short root vegetables. The use of hydroponic crops (or other soilless production) also enables agricultural production in degraded urban areas [50], [51].

In developing countries, a considerable portion of the population lives in slum areas, where the lack of land regularization impedes the development of sustainable agriculture. Lack of a sense of ownership leads to the development of a predatory and low-investment type of agriculture onsite, with no concern for soil or water protection techniques that ensure the continuation of farming activity over time [52].

Urban and peri-urban spaces are affected by the deposition of residential waste. Indicators reflecting the percentage of the population served by regular waste collection services and the recycling percentage may show whether a community is reducing its production of waste. These are measures that favour sustainable development. Yates and Gutberlet [53] recall that use of the organic fraction of urban waste generates the recirculation of nutrients and reduces the production of methane, a greenhouse gas.

Finally, another important input for sustainable UA would be an atmosphere free of contaminating agents. Amato-Lourenço et al. [54] detected air pollutants deposited on the leaves of vegetables grown in an urban environment and showed that higher concentrations of heavy metals were correlated with the presence of traffic in the surrounding area.

4 CONCLUSIONS

The 26 indicators proposed to compose an index for the assessment of UA sustainability were inspired by the scientific and technical literature on urban sustainability. These indicators were grouped according to five categories of analysis, and also separated into sub-categories. These categories were defined according to the dimension of the food-energy-water-land/space-air pollution Nexus. They all met the criteria of being measurable, sensitive to stress, predictable, anticipatory, controllable, integrative, responsive and stable.

The index should show, on a scale ranging from zero to one, the degree to which the UA practiced by a city fulfils the sustainability criteria and targets set by 2030 UN SDGs. The next step in this research will involve validation of these indicators and determination of the weight of each indicator, to be defined via an Analytic Hierarchy Process, with the support of specialists, according to the criteria of the Delphi methodology. Through the attribution of weights, normalization and distance in relation to pre-established targets, improvements or reductions in the set of 26 indicators could support decision makers and public administrators in identifying and implementing corrective actions towards sustainable development.

REFERENCES

[1]   Sanyé-Mengual, E., *COST Action Urban Agriculture Europe: Freelance STSM: Stakeholders' Acceptance, Governance and Power Relations in Innovative Forms of Urban Agriculture*, Muncheberg, Germany, 2015. https://doi.org/https://doi.org/10.13140/RG.2.1.2659.4966.

[2]   UN DESA, Sustainable development goals report 2018. United Nations Department of Economic and Social Affair: New York, 2018.

[3]   Hoff, H., Understanding the Nexus. Background paper for the *Bonn2011 Nexus Conference*, Stockholm Environment Institute, (Nov.), pp. 1–52, 2011.

[4]   Allouche, J., Middleton, C. & Gyawali, D., Technical veil, hidden politics: Interrogating the power linkages behind the nexus. *Water Alternatives*, **8**(1), pp. 610–626, 2015.

[5]   Pintér, L., Hardi, P., Martinuzzi, A. & Hall, J., Bellagio STAMP: Principles for sustainability assessment and measurement. *Ecological Indicators*, **17**, pp. 20–28, 2012. https://doi.org/10.1016/j.ecolind.2011.07.001.

[6]   Turnhout, E., Hisschemöller, M. & Eijsackers, H., Ecological indicators: Between the two fires of science and policy. *Ecological Indicators*, **7**(2), pp. 215–228, 2007. https://doi.org/10.1016/j.ecolind.2005.12.003.

[7]   Dizdaroglu, D., Developing micro-level urban ecosystem indicators for sustainability assessment. *Environmental Impact Assessment Review*, **54**, pp. 119–124, 2015. https://doi.org/10.1016/j.eiar.2015.06.004.

[8]   Peng, J., Liu, Z., Liu, Y., Hu, X. & Wang, A., Multifunctionality assessment of urban agriculture in Beijing City, China. *Science of the Total Environment*, **537**, pp. 343–351, 2015. https://doi.org/10.1016/j.scitotenv.2015.07.136.

[9]   Grewal, S.S. & Grewal, P.S., Can cities become self-reliant in food? *Cities*, **29**(1), pp. 1–11, 2012. https://doi.org/10.1016/j.cities.2011.06.003.

[10] Saha, M. & Eckelman, M.J., Growing fresh fruits and vegetables in an urban landscape: A geospatial assessment of ground level and rooftop urban agriculture potential in Boston, USA. *Landscape and Urban Planning*, **165**(August 2016), pp. 130–141, 2017. https://doi.org/10.1016/j.landurbplan.2017.04.015.

[11] Orsini, F. et al., Exploring the production capacity of rooftop gardens (RTGs) in urban agriculture: the potential impact on food and nutrition security, biodiversity and other ecosystem services in the city of Bologna. *Food Security*, **6**(6), pp. 781–792, 2014. https://doi.org/10.1007/s12571-014-0389-6.

[12] European Commission, European Commission's Directorate-General Environment, Depth report: Indicators for sustainable cities. UWE: Bristol, 2015. https://doi.org/10.2779/61700.

[13] Santini, G., Miller, S. & Dubbeling, M., *City Region Food System Tools/Examples City Region Food System Tools/Examples*, FAO: Rome, 2018. www.fao.org/documents/card/en/c/I9255EN. Accessed on: 21 Feb. 2020.

[14]   FAO/MUFPP/RUAF, *The Milan Urban Food Policy Pact Monitoring Framework*, Rome, 2019. www.milanurbanfoodpolicypact.org/wp-content/uploads/2019/11/CA6144EN.pdf. Accessed on: 21 Feb. 2020.

[15]   Zurek, M. et al., Sustainability metrics for the EU food system: A review across economic, environmental and social considerations, 2017. https://susfans.eu/system/files/public_files/Publications/Reports/SUSFANS-Deliverable–D1.3-UOXF.pdf. Accessed on: 21 Feb. 2020.

[16]   van Leeuwen, C.J., Frijns, J., van Wezel, A. & van de Ven, F.H.M., City Blueprints: 24 indicators to assess the sustainability of the urban water cycle. *Water Resources Management*, **26**(8), pp. 2177–2197, 2012. https://doi.org/10.1007/s11269-012-0009-1.

[17]   Koop, S.H.A. & van Leeuwen, C.J., Assessment of the sustainability of water resources management: A critical review of the city blueprint approach. *Water Resources Management*, **29**(15), pp. 5649–5670, 2015. https://doi.org/10.1007/s11269-015-1139-z.

[18]   Economist Intelligence Unit, European Green City Index. Siemens AG: Munich, p. 51, 2009.

[19]   University of Toronto, Global cities. List of indicators: Global city indicators facility 2007. http://www.globalcitiesinstitute.org. Accessed on: 21 Feb. 2020.

[20]   ISO/TC 268, ISO 37120 – Sustainable development of communities, Indicators for city services and quality of life, 2014.

[21]   Canadian International Development Agency (CIDA), Indicators for Sustainability – How cities are monitoring and evaluating their success. Canada, 2012.

[22]   Van Dijken, K., Dorenbos, R. & Kamphof, R., *The Reference Framework for Sustainable Cities (RFSC) Testing Results and Recommendations*, The Hague, The Netherlands, 2012. https://www.eukn.eu/fileadmin/Lib/files/EUKN/2013/Finalreport NicistestingRFSC.pdf. Accessed on: 21 Feb. 2020.

[23]   Star Community, Sustainability Tools for Assessing and Rating Communities (STAR) Community Rating System Version 2.0. http://www.starcommunities.org. Accessed on: 21 Feb. 2020.

[24]   Berrini, M. & Bono, L., Report 2007 urban ecosystem Europe: An integrated assessment on the sustainability of 32 European cities. Milan, Italy, 2007.

[25]   Minx, J., Creutzig, F., Ziegler, T. & Owen, A., Developing a pragmatic approach to assess urban metabolism in Europe – A report to the Environment Agency prepared by Technische Universität Berlin and Stockholm Environment Institute, Climatecon Working Paper 01/2011, Technische Universität Berlin: Berlin, Vol. 240, 2011.

[26]   Mega, V. & Pedersen, J., *Urban Sustainability Indicators. European Foundation*, Luxembourg, 1998. https://doi.org/10.18848/1832-2077/cgp/v07i06/55007.

[27]   UN-Habitat, Urban Indicators Guidelines – UN-Habitat, 2004.

[28]   Zhang, P. et al., Food-energy-water (FEW) nexus for urban sustainability: A comprehensive review. *Resources, Conservation and Recycling*, **142**(Nov. 2018), pp. 215–224, 2019. https://doi.org/10.1016/j.resconrec.2018.11.018.

[29]   Ringler, C., Bhaduri, A. & Lawford, R., The nexus across water, energy, land and food (WELF): Potential for improved resource use efficiency? *Current Opinion in Environmental Sustainability*, **5**(6), pp. 617–624, 2013. https://doi.org/10.1016/j.cosust.2013.11.002.

[30]   Karabulut, A.A., Crenna, E., Sala, S. & Udias, A., A proposal for integration of the ecosystem-water-food-land-energy (EWFLE) nexus concept into life cycle assessment: A synthesis matrix system for food security. *Journal of Cleaner Production*, **172**, pp. 3874–3889, 2018. https://doi.org/10.1016/j.jclepro.2017.05.092.

[31]   Kumar, P. & Saroj, D.P., Water-energy-pollution nexus for growing cities. *Urban Climate*, **10**(P5), pp. 846–853, 2014. https://doi.org/10.1016/j.uclim.2014.07.004.

[32]   Dale, V.H. & Beyeler, S.C., Challenges in the development and use of ecological indicators. *Ecological Indicators*, **1**(1), pp. 3–10, 2001. https://doi.org/10.1016/S1470-160X(01)00003-6.

[33]   Lin, T., Lin, J., Cui, Y., Hui, S. & Cameron, S., Using a network framework to quantitatively select ecological indicators. *Ecological Indicators*, **9**(6), pp. 1114–1120, 2009. https://doi.org/10.1016/j.ecolind.2008.12.009.

[34]   Willett, W. et al., Food in the Anthropocene: the EAT–Lancet Commission on healthy diets from sustainable food systems. *Lancet*, **6736**, pp. 3–49, 2019. https://doi.org/10.1016/S0140-6736(18)31788-4.

[35]   Moran, D.D., Wackernagel, M., Kitzes, J.A., Goldfinger, S.H. & Boutaud, A., Measuring sustainable development – Nation by nation. *Ecological Economics*, **64**(3), pp. 470–474, 2008. https://doi.org/10.1016/j.ecolecon.2007.08.017.

[36]   Orsini, F., Kahane, R., Nono-Womdim, R. & Gianquinto, G., Urban agriculture in the developing world: A review. *Agronomy for Sustainable Development*, **33**(4), pp. 695–720, 2013. https://doi.org/10.1007/s13593-013-0143-z.

[37]   Song, T., Cai, J., Chahine, T. & Li, L., Towards smart cities by Internet of Things (IoT) – a Silent Revolution in China. *Journal of the Knowledge Economy*, 2017. https://doi.org/https://doi.org/10.1007/s13132-017-0493-x.

[38]   Nchanji, E.B. et al., Assessing the sustainability of vegetable production practices in northern Ghana*. *International Journal of Agricultural Sustainability*, **15**(3), pp. 321–337, 2017. https://doi.org/10.1080/14735903.2017.1312796.

[39]   Gerster-Bentaya, M., Nutrition-sensitive urban agriculture. *Food Security*, **5**(5), pp. 723–737, 2013. https://doi.org/10.1007/s12571-013-0295-3.

[40]   Borges, C.A., Cabral-Miranda, W. & Jaime, P.C., Urban food sources and the challenges of food availability according to the Brazilian dietary guidelines recommendations. *Sustainability (Switzerland)*, **10**(12), 2018. https://doi.org/10.3390/su10124643.

[41]   Schnell, S.M., Food miles, local eating, and community supported agriculture: Putting local food in its place. *Agriculture and Human Values*, **30**(4), pp. 615–628, 2013. https://doi.org/10.1007/s10460-013-9436-8.

[42]   Eigenbrod, C. & Gruda, N., Urban vegetable for food security in cities. A review. *Agronomy for Sustainable Development*, **35**(2), pp. 483–498, 2015. https://doi.org/10.1007/s13593-014-0273-y

[43]   Goldstein, B., Hauschild, M., Fernández, J. & Birkved, M., Testing the environmental performance of urban agriculture as a food supply in northern climates. *Journal of Cleaner Production*, **135**, pp. 984–994, 2016. https://doi.org/10.1016/j.jclepro.2016.07.004.

[44]   Riahi, K. et al., The Shared Socioeconomic Pathways and their energy, land use, and greenhouse gas emissions implications: An overview. *Global Environmental Change*, **42**, pp. 153–168, 2017. https://doi.org/10.1016/j.gloenvcha.2016.05.009.

[45] IPCC, Summary for policymakers. *IPCC Fifth Assessment Report: Working Group III Mitigation of Climate Change*, eds T.Z. Edenhofer et al., Cambridge University Press: Cambridge, UK, and New York, pp. 1–33, 2014. https://doi.org/10.1017/CBO9781107415324

[46] Lupia, F., Baiocchi, V., Lelo, K. & Pulighe, G., Exploring rooftop rainwater harvesting potential for food production in urban areas. *Agriculture (Switzerland)*, **7**(6), 2017. https://doi.org/10.3390/agriculture7060046.

[47] Paviotti-Fischer, E., Lopes-Torres, E.J., Santos, M.A.J., Brandolini, S.V.P.B. & Pinheiro, J., Xiphidiocercariae from naturally infected Lymnaea columella (Mollusca, Gastropoda) in urban area: Morphology and ultrastructure of the larvae and histological changes in the mollusc host. *Brazilian Journal of Biology*, **79**(3), pp. 446–451, 2019. https://doi.org/10.1590/1519-6984.182501.

[48] Kumar, K., & Hundal, L.S., Soil in the city: Sustainably improving urban soils. *Journal of Environmental Quality*, **45**(1), pp. 2–8, 2016. https://doi.org/10.2134/jeq2015.11.0589.

[49] Lin, B.B., Philpott, S.M. & Jha, S., The future of urban agriculture and biodiversity-ecosystem services: Challenges and next steps. *Basic and Applied Ecology*, **16**(3), pp. 189–201, 2015. https://doi.org/10.1016/j.baae.2015.01.005.

[50] Brown, K.H. & Jameton, A.L., Public health implications of urban agriculture. *Journal of Public Health Policy*, **21**(1), pp. 20–39, 2000. https://doi.org/10.2307/3343472.

[51] Brown, S.L., Chaney, R.L. & Hettiarachchi, G.M., Lead in urban soils: A real or perceived concern for urban agriculture? *Journal of Environmental Quality*, **45**(1), pp. 26–36, 2016. https://doi.org/10.2134/jeq2015.07.0376.

[52] Drechsel, P. & Dongus, S., Dynamics and sustainability of urban agriculture: Examples from sub-Saharan Africa. *Sustainability Science*, **5**(1), pp. 69–78, 2010. https://doi.org/10.1007/s11625-009-0097-x.

[53] Yates, J.S. & Gutberlet, J., Reclaiming and recirculating urban natures: Integrated organic waste management in Diadema, Brazil. *Environment and Planning A*, **43**(9), pp. 2109–2124, 2011. https://doi.org/10.1068/a4439.

[54] Amato-Lourenco, L.F. et al., The influence of atmospheric particles on the elemental content of vegetables in urban gardens of Sao Paulo, Brazil. *Environmental Pollution*, **216**, pp. 125–134, 2016. https://doi.org/10.1016/j.envpol.2016.05.036.

SECTION 4
URBAN METABOLISM

GREEN ROOF PERFORMANCE IN SUSTAINABLE CITIES

ANITA RAIMONDI[1], GIANFRANCO BECCIU[1], UMBERTO SANFILIPPO[1] & STEFANO MAMBRETTI[1,2]
[1]Politecnico di Milano, Italy
[2]Wessex Institute of Technology, UK

ABSTRACT

In the last few decades, the use of sustainable urban drainage systems is largely spreading and encouraged, because they provide lots of benefits for sewer networks, wastewater treatment plants and the environment. In this context, green roofs can be an effective tool to both delay and attenuate stormwater runoff peaks, reducing runoff at the same time. Their proper design is a key element for stormwater management in highly urbanized cities. The aim of this paper is to propose an analytical probabilistic approach, to evaluate green roof performance in terms of runoff and vegetation's survival without irrigation, to guide planners in choosing proper values for their design parameters. A great advantage of the method is that it can be applied to different sites and climate conditions; moreover, it involves a significant improvement of the typical analytical probabilistic approach, as a chain of consecutive rainfall events was considered, in order to take into account the possibility that storage capacity is not completely available at the beginning of each event, because of pre-filling from previous rainfalls, as typically happens with green roofs. Finally, to verify the goodness of our developed equations, we applied them to a case study.
Keywords: construction, sustainable urban drainage systems, green roof, green architecture, analytical probabilistic modelling, vegetation survival, runoff, stormwater runoff, sustainability.

1 INTRODUCTION

The rapid growth of the urban world's population and the consequent increase of highly urbanised areas, joined to climate changes underway involves a strong imbalance of the natural water cycle with frequent flooding and surface water contamination. In this context, the implementation of sustainable urban drainage systems (SUDSs) can deliver several benefits: relief of the loads in sewer networks, increased efficiency of wastewater treatments plants, reduction of polluted waters discharged into the environment and the increase of biodiversity in urban areas [1]. Among these strategies, green roofs can be an effective countermeasure, because their implementation also entails significant environmental and economic benefits; besides stormwater management and the improvement of quality of receiving water bodies, such as energy savings, the reduction of heat island effects and the enhancement of biodiversity. Moreover, they do not require additional space with respect to the building footprint and can be an effective tool in densely built urban areas, where rooftops constitute about 30–50% of urban impermeable surfaces [2]. With reference to stormwater management, green roofs allow the local disposal of rainwater runoff, the reduction of runoff volumes through evapotranspiration from the vegetation and exposed surfaces, the delay of runoff that starts only after soil saturation, the reduction and delay of runoff peak rates for the infiltration of rainwater into the soil, their temporary storage in the substrate and drainage layer, and the improvement of stormwater quality for effective percolation into the soil.

The first green roofs appeared in Germany in the 1970s [3]; since then, they have spread all over the world, especially in the major modern and advanced cities, where incentive programs were often developed to encourage or even impose their installation. In the last decades, several studies and models deepened green roof performance under different climate and design conditions [4]–[8], often with reference to a specific place or climate [9]–[12] on a small spatial and time scale [13], [14]. At present, there is not yet enough scientific evidence to demonstrate the hydrological benefits of green roofs for full-scale installations [15], [16].

The aim of this paper is to guide planners in the design of green roofs for the achievement of a double goal: to guarantee an appropriate water retention capacity and vegetation survival without watering. The use of an analytical probabilistic approach is proposed. This kind of methodology, first proposed by Guo and Adams [17]–[19] and Adams and Papa [20], allows the estimation of the probability distribution functions (PDFs) of the variables of interest from the PDF of input rainfall variables, once the analytical relation and the variables characterizing the process are defined. One of the main advantages of the analytical probabilistic approaches is that of combined simplicity of "design storm" methods and the probabilistic reliability of continuous simulations. Moreover, the resulting equations are not calibrated on a specific place, but can be applied to different basins and climates all over the world. In recent years, the analytical probabilistic approach has been applied by different authors also to SUDS such as rain water harvesting systems [21]–[24], infiltration trenches [25], [26], permeable pavements [27], bioretention systems [28], green roofs [29]–[31] and stormwater detention facilities [32]–[37]. In this paper, the PDF of runoff from green roofs and of water content into growing medium at the end of dry periods have been estimated, with the dual purpose of driving planners in towards the choice of best green roof layer thickness, to guarantee a good retention capacity and survival of vegetation without irrigation.

An important innovation of our proposed manuscript is that it allows us to consider pre-filling from a chain of previous rainfall events, that gives a possibility that the retention capacity is not completely available at the beginning of the considered rainfall. This aspect is particularly relevant for SUDS like green roofs, characterised by low outflow rates, limited to the degree of evapotranspiration from vegetation and soil [32], [35]. The method allows an optimum green roof design, as the resulting expressions relate the growing medium thickness to the average return interval (ARI). To validate our developed equations, an application to a case study in Milano, Italy is proposed.

2 HYDROLOGICAL MODELLING

Green roofs are engineered multi-layered structures, with a vegetated upper surface, working in shallow systems without connection to natural ground. A typical green roof is composed of the vegetation layer, the growing media layer, a blend of mineral material enriched with organic material where water is retained and in which vegetation is anchored, the filter fabric; the drainage layer, generally constituted of plastic profiled elements that store water for the plants' sustainment during dry periods, evacuating excess water in roof drains; the root resistant membrane and the mechanical protection geotextile. Fig. 1 shows the conceptual reference scheme considered in this paper for green roof modelling.

The volumes in Fig. 1 must be intended as specific for the unit area. Rainfall is first intercepted by the vegetation, then infiltrated into the growing medium layer, where it is retained, used by the roots and released back into the atmosphere through evapotranspiration. The excess is stored in the underlying drainage layer, equipped with an overflow to the urban drainage system that activates when the retention capacity is filled.

Green roof design includes the selection of the thickness of each layer and the choice of plant cover. Rainwater stored in a green roof can vary between zero, when it is completely dry and w_{max} if the water content in the green roof is at its maximum, as in the case of two very close heavy rainfall events. Interception by vegetation generally equals just a few millimetres and the drainage layer capacity generally is between 5–10 cm; the focus here is

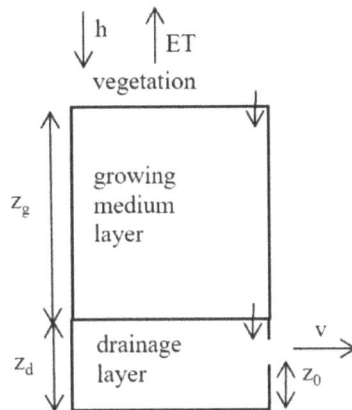

Figure 1:   Green roof conceptual scheme considered for our modelling. h: rainfall depth; ET: evapotranspiration; v: runoff; z_g: growing medium thickness; z_d: drainage layer thickness; z_0: overflow height.

on the growing medium layer that can vary between zero and $\phi_f \cdot z_g$, where ϕ_f represents the growing medium moisture content at saturation, when the storage volume is full. In the estimation of the probability of runoff from green roofs, the term w_{max} refers to the sum of the maximum retention capacity in the three layers (vegetation, growing medium and drainage), while in the estimation of the probability of having a minimum water content enough to allow vegetation survival without irrigation, only the maximum retention capacity of the growing medium is considered. Extended rainless periods, especially occurring during the hot season, can result in the soil moisture falling to a "wilting point", with subsequent death of the plant cover. The death of the vegetation nullifies transpiration, but progressive soil desiccation, on the other hand, is initially in some way positive, in terms of increased capacity to buffer runoff.

The evapotranspiration rate, that is, the amount of water released to the atmosphere from the plants' transpiration and soil evaporation, strictly depends on rainfall depth and the water content of the growing medium from previous rainfalls. In the modelling, it is assumed always equal to potential evapotranspiration, that is its maximum value, the worst condition for vegetation.

Water content in the growing medium is estimated at the end of the rainfall event (subscript u) for the estimation of the probability of runoff and at the end of the dry period (subscript e) for the estimation of the probability to have a minimum water content that would allow vegetation survival without irrigation. Considering a variable number $i = 1, ..., N$ of chained rainfall events, final water depth in the growing medium is calculated by eqn (1):

$$h_{gu,i-1} = \begin{cases} h_{ge,i-1} + h_{i-1} - Et \cdot \theta_{i-1} & 0 \leq h_{ge,i-1} + h_{i-1} - Et \cdot \theta_{i-1} < \phi_f \cdot z_g \\ \phi_f \cdot z_g & h_{ge,i-1} + h_{i-1} - Et \cdot \theta_{i-1} \geq \phi_f \cdot z_g \\ 0 & otherwise. \end{cases} \quad (1)$$

For $i = 1$, the growing medium is considered empty at the beginning of the event ($h_{ge,0} = 0$). Water depth in the growing medium at the beginning of a generic rainfall (always considering $i = 1, ..., N$) results in eqn (2):

$$h_{ge,i} = \begin{cases} h_{gu,i-1} - Et \cdot d_{i-1} & h_{gu,i-1} - Et \cdot d_{i-1} > 0 \\ 0 & otherwise. \end{cases} \tag{2}$$

Runoff at the end of a generic event v_i is expressed by eqn (3):

$$v_i = \begin{cases} h_{gu,i-1} - Et \cdot d_i + h_i - Et \cdot \theta_i - w_{max} & Condition_1 \\ h_i - Et \cdot \theta_i - w_{max} & Condition_2; Condition_3 \\ w_{max} - Et \cdot d_i + h_i - Et \cdot \theta_i - w_{max} & Condition_4 \\ 0 & Otherwise, \end{cases} \tag{3}$$

$Condition_1$: $h_{gu,i-1} \leq w_{max}$; $h_{gu,i-1} > Et \cdot d_i$; $h_{gu,i-1} - Et \cdot d_i + h_i - Et \cdot \theta_i > w_{max}$,

$Condition_2$: $h_{gu,i-1} \leq w_{max}$; $h_{gu,i-1} \leq Et \cdot d_i$; $h_i - Et \cdot \theta_i > w_{max}$,

$Condition_3$: $h_{gu,i-1} > w_{max}$; $w_{max} \leq Et \cdot d_i$; $h_i - Et \cdot \theta_i > w_{max}$,

$Condition_4$: $h_{gu,i-1} > w_{max}$; $w_{max} > Et \cdot d_i$; $w_{max} - Et \cdot d_i + h_i - Et \cdot \theta_i > w_{max}$.

For $i = 0$, that is v_0, runoff results, as seen in eqn (4):

$$v_0 = \begin{cases} h_0 - Et \cdot \theta_0 - w_{max} & h_i - Et \cdot \theta_i > w_{max} \\ 0 & Otherwise. \end{cases} \tag{4}$$

With reference to eqn (4):

- Condition 1 expresses the case that there is no runoff at the end of event $i - 1$, there is pre-filling from event $i - 1$ at the beginning of event i and there is runoff from the green roof at the end of event i;
- Condition 2 expresses the case that there is no runoff at the end of event $i - 1$, there is no pre-filling from event $i - 1$ at the beginning of event i and there is runoff from the green roof at the end of event i;
- Condition 3 expresses the case that there is runoff at the end of event $i - 1$, there is no pre-filling from event $i - 1$ at the beginning of event i and there is runoff from the green roof at the end of event i;
- Condition 4 expresses the case that there is runoff at the end of event $i - 1$, there is pre-filling from event $i - 1$ at the beginning of event i and there is runoff from the green roof at the end of event i.

3 PROBABILISTIC MODEL

The aim of the proposed probabilistic model is to give an estimate of the growing medium thickness to be considered in green roof designs, in order to have, with an assumed probability, a minimum water content enough to allow vegetation survival without irrigation and limited runoff. Hydrological variables considered in our modelling are: rainfall depth h, rainfall duration θ and inter-event time d; these are assumed to be independent and exponentially distributed rainfall variables. To isolate independent rainfalls from a continuous record of events, a minimum inter-event time (IETD) was defined [38]. If the inter-event time between two consecutive rainfalls was smaller than IETD, they were joined into a single event, otherwise they were considered as distinct and independent.

The assumption of the exponential PDF for rainfall variables is usually considered acceptable to satisfy the need of an easier mathematical tractability [39]–[41]. To overcome the bias due to the use of the exponential PDF, some authors suggest using the Weibull PDF [42] or the double-exponential PDF [34]. Although a better fitting of the observed frequencies of rainfall records can be achieved with these alternative PDFs, the improvement in terms of model accuracy seems negligible, compared with the significant increase of the mathematical complexity in the development of equations.

Exponential PDFs of rainfall depth, rainfall duration and inter-event time are expressed as, respectively, eqns (5)–(7):

$$f_h = \xi \cdot e^{-\xi \cdot h}, \tag{5}$$

$$f_\theta = \lambda \cdot e^{-\lambda \cdot \theta}, \tag{6}$$

$$f_d = \psi \cdot e^{-\psi \cdot (d - IETD)}, \tag{7}$$

where $\xi = 1/\mu_h$; $\lambda = 1/\mu_\theta$; $\psi = 1/(\mu_d - IETD)$ and μ_h, μ_θ and μ_d are respectively the mean values of rainfall depth, rainfall duration and interevent time.

The probability of having a runoff exceeding a given threshold $\bar{v}$ and the probability that water content in the growing medium exceeds a minimum threshold w is estimated, setting $h = h_i = h_{i+1}$, $\theta = \theta_i = \theta_{i+1}$, $d = d_i = d_{i+1}$ in eqns (1)–(4): this involves the deletion of Condition 2. Runoff PDF is estimated to distinguish two different conditions: maximum emptying time, that is the time needed to empty the retention capacity when it is full, lower (Case 1) and higher (Case 2) than $IETD$; for Case 1 pre-filling from previous rainfalls is excluded and the full storage capacity is available, while for Case 2 the possibility that the retention volume is partially filled from previous rainfalls was considered.

Case 1: $w_{max}/Et \leq IETD$:

$$P_{v1} = P(v > \bar{v}) = \int_{h=w_{max}+\bar{v}+Et\cdot\theta}^{\infty} f_h \cdot dh \int_{\theta=0}^{\infty} f_\theta \cdot d\theta = \gamma \cdot e^{-\xi \cdot (w_{max}+\bar{v})}, \tag{8}$$

where $\gamma = \dfrac{\lambda}{\lambda + Et \cdot \xi}$.

Case 2: $w_{max}/Et > IETD$:

$$P_{v2} = P(v > \bar{v}) = \int_{\theta=0}^{\infty} f_\theta \cdot d\theta \int_{d=IETD}^{\infty} f_d \cdot dd \int_{h=w_{max}+\bar{v}+Et\cdot\theta}^{\infty} f_h \cdot dh$$

$$+ \sum_{i=2}^{N} \left[\int_{\theta=0}^{\infty} f_\theta \cdot d\theta \int_{d=IETD}^{\frac{w_{max}+\bar{v}}{Et}} f_d \cdot dd \int_{h=\frac{w_{max}+\bar{v}+(i-1)\cdot Et\cdot d}{i}+Et\cdot\theta}^{\frac{w_{max}+\bar{v}+(i-2)\cdot Et\cdot d}{i-1}+Et\cdot\theta} f_h \cdot dh \right]$$

$$= \gamma \cdot \left\{ e^{-\xi \cdot (w_{max}+\bar{v})} + \psi \cdot \sum_{i=2}^{N} \left[-(i-1)\cdot\beta_i \cdot e^{-\xi \cdot Et \cdot IETD \cdot \left(\frac{i-2}{i-1}\right) - \frac{\xi}{i-1}(\bar{v}+w_{max})} - i \cdot \beta_i^* \cdot \right. \right.$$

$$\left. \left. e^{-\frac{\xi}{i}[Et \cdot IETD \cdot (i-1) + (\bar{v}+w_{max})]} - \xi \cdot Et \cdot \beta_i \cdot \beta_i^* \cdot e^{\psi \cdot IETD - (\bar{v}+w_{max})\cdot\left(\frac{\psi}{Et}+\xi\right)} \right] \right\}, \tag{9}$$

with: $\beta_i = \dfrac{1}{\xi \cdot Et \cdot (i-2) + \psi \cdot (i-1)}$; $\beta_i^* = -\dfrac{1}{i \cdot \psi + (i-1) \cdot \xi \cdot Et}$.

To estimate the probability for green roof vegetation to survive without irrigation, a minimum water content w is considered. The condition for which water content can be different from zero at the end of a dry period, that is when pre-filling from previous events is considered, results to be: $(z_g \cdot \phi_f - w)/Et > IETD_g \cdot \phi_f$.

Two different cases are analyzed: a single rainfall ($i = 1$) and a series of chained (N) rainfall events ($i > 1$).

For $i = 1$, that is when a single rainfall event is considered, the exceedance probability to have, at the end of the dry period between two consecutive rainfalls, a minimum water volume in the growing medium, to ensure vegetation survival, results in eqn (10):

$$P_{w1} = P(h_{ge} > w) = \int_{h=w+Et\cdot(d+\theta)}^{\infty} f_h \cdot dh \int_{d=IETD}^{\frac{\emptyset_f \cdot z_g - w}{Et}} f_d \cdot dd \int_{\theta=0}^{\infty} f_\theta \cdot d\theta$$

$$= \gamma \cdot \beta \cdot \left[e^{-\xi \cdot (Et \cdot IETD + w)} - e^{\psi \cdot \left(IETD + \frac{w}{Et}\right) - \emptyset_f \cdot z_g \cdot \left(\xi + \frac{\psi}{Et}\right)} \right]. \qquad (10)$$

For $i > 1$, that is if a series of chained rainfalls is considered, the probability was as given by eqn (11):

$$P_{wN} = P(h_{ge} > w)$$

$$= \int_{\theta=0}^{\infty} f_\theta \cdot d\theta$$

$$\cdot \left\{ \int_{d=IETD}^{\frac{\emptyset_f \cdot z_g - w}{Et}} f_d \cdot dd \cdot \left[\int_{h=\emptyset_f \cdot z_g + Et \cdot \theta}^{\infty} f_h \cdot dh \cdot + \int_{h=\frac{w}{i-1}+Et\cdot(\theta+d)}^{\frac{\emptyset_f \cdot z_g + Et \cdot d \cdot (i-2)}{i-1}+Et\cdot\theta} f_h \cdot dh \right] \right.$$

$$\left. + \int_{\frac{\emptyset_f \cdot z_g + Et \cdot d \cdot (i-1)}{i}+Et\cdot\theta}^{\frac{\emptyset_f \cdot z_g + Et \cdot d \cdot (i-2)}{i-1}+Et\cdot\theta} f_h \cdot dh \int_{d=IETD}^{d=\frac{\emptyset_f \cdot z_g \cdot (i-1)-i\cdot w}{Et\cdot(i-1)}} f_d \cdot dd \right\} =$$

$$= \gamma \cdot \left\{ e^{-\xi \cdot \emptyset_f \cdot z_g} \cdot \left[1 - e^{\psi \cdot \left(IETD + \frac{w}{Et} - \frac{\emptyset_f \cdot z_g}{Et}\right)} \right] + \frac{2 \cdot (1-\beta) \cdot \beta_i}{i-1} \cdot e^{-\left(\frac{\psi}{Et}+\xi\right) \cdot (\emptyset_f \cdot z_g - w) + \psi \cdot IETD - \frac{\xi \cdot w}{i-1}} - \right.$$

$$\beta_i \left[2 \cdot e^{-\frac{\xi}{i-1}[\emptyset_f \cdot z_g + Et \cdot IETD \cdot (i-2)]} + e^{\frac{\xi \cdot w \cdot i \cdot (i-2)}{(i-1)^2} - \emptyset_f \cdot z_g \cdot \left(\frac{\psi}{Et}+\xi\right) + \psi \cdot IETD + \frac{\psi \cdot i \cdot w}{Et \cdot (i-1)}} \right] + \beta \cdot$$

$$\left. e^{-\frac{\xi \cdot w}{i-1} - \xi \cdot Et \cdot IETD} + \beta_i^* \cdot \left[e^{-\emptyset_f \cdot z_g \cdot \left(\frac{\psi}{Et}+\xi\right) + \psi \cdot IETD + \frac{\psi \cdot i \cdot w}{Et \cdot (i-1)}} - e^{-\frac{\xi}{i}[\emptyset_f \cdot z_g + Et \cdot IETD \cdot (i-1)]} \right] \right\}. \qquad (11)$$

The quantities γ, β, β_i and β_i^* are equal to:

$$\gamma = \frac{\lambda}{\lambda + \xi \cdot Et},$$

$$\beta = \frac{\psi}{\psi + \xi \cdot Et},$$

$$\beta_i = \frac{(i-1) \cdot \psi}{(i-1) \cdot \psi + \xi \cdot Et \cdot (i-2)},$$

$$\beta_i^* = \frac{\psi \cdot i}{\xi \cdot Et - i \cdot (\xi \cdot Et + \psi)}.$$

Eqns (10) and (11) can be used to estimate the growing medium thickness z_g required for a green roof design, once variables are defined characterising rainfall, vegetation and exceedance probability.

4 APPLICATION TO A CASE STUDY

Eqns (8)–(11) were applied to a case study in Milan, Italy, using statistics from the series of rainfall events recorded at the Milano-Monviso gauge station in the period 1971–2005. To assess the PDFs of runoff and water content for growing medium, an IETD = 10 hours was assumed. Table 1 reports the mean and coefficient of variation of rainfall depth h, rainfall duration θ and inter-event time d.

Table 1: Mean and coefficient of variation of rainfall variables.

	IETD = 10 hours	
	μ (mm)	V (–)
h (mm)	18.49	1.15
θ (hours)	14.37	1.03
d (hours)	172.81	1.30

As already discussed [35], the assumption of exponential distribution for rainfall variables is not the best fitting choice for these data. Table 2 reports the correlation indexes among rainfall variables.

Table 2: Correlation indexes among rainfall variables.

	IETD = 10 hours
$\rho_{h,d}$ (–)	0.11
$\rho_{\theta,h}$ (–)	0.62
$\rho_{d,\theta}$ (–)	0.11

Inter-event time results to be only weakly correlated to other two variables, while correlation between rainfall depth and duration was significant. The effects of this correlation on final results have been already discussed by the authors [33].

Runoff probability P_v is estimated by varying maximum retention capacity w_{max}. It results from the sum of the maximum retention capacity of vegetation, growing medium and drainage layer. Maximum retention capacity of vegetation is of few millimetres; maximum retention capacity in growing medium can generally vary between 0 and 1,000 mm, considering both extensive green roofs (with a thickness of the layer equal to few centimetres) and intensive green roofs (with a thickness of the layer of some hundreds of centimetres; the maximum retention capacity of the drainage layer usually varies from 0–150 mm, so that the maximum retention capacity of the whole roof can vary between 0 and 1,250 mm.

In the calculation, an extensive green roof has been considered and the maximum retention capacity was varied between 0 and 200 mm. Water content at saturation is assumed equal to $\phi_f = 0.58$ (–) [43]. The growing medium thickness z_g is assumed variable between 20 mm and 500 mm. Different studies in the literature collected results of field measurements, trying to define a range of reasonable values of green roof evapotranspiration rates: experimental estimations range from 0.69 to 6–9 mm/day, with typical values of 1–6 mm/day [44], [45]. In this paper, a value of evapotranspiration rate equal to Et=0.125 mm/hour was used. In the calculation of both probability P_v and P_w threshold values $\bar{v}$ and w were set equal to zero. Exceedance probabilities calculated by eqns (8)–(11) are compared with cumulative frequencies obtained by the continuous simulation of recorded rainfalls; the number of considered chain events to achieve a good fit between P and F is equal to i = 5.

WIT Transactions on Ecology and the Environment, Vol 243, © 2020 WIT Press
www.witpress.com, ISSN 1743-3541 (on-line)

Fig. 2 shows the water content PDF for vegetation survival without irrigation; the probability increases with the thickness of the growing medium.

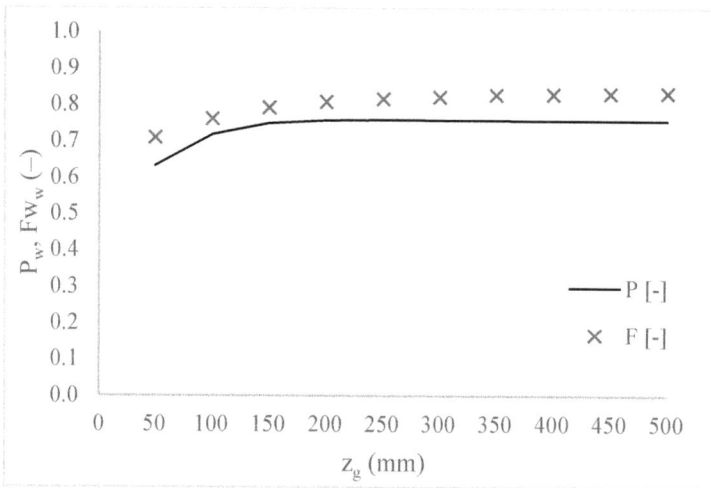

Figure 2:   Probability (P) and frequency (F) distribution functions of water content into growing medium layer varying its thickness (z_g).

Table 3: Analysis results on the whole period of records.

z_g (mm)	50	100	150	200	250	300	350	400	450	500
P (mm)	0.63	0.72	0.75	0.76	0.76	0.76	0.76	0.76	0.76	0.76
T (years)	3	4	4	4	4	4	4	4	4	4

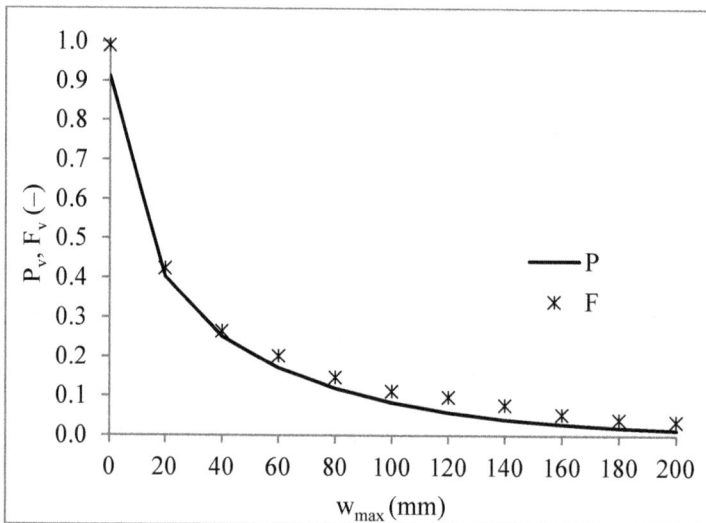

Figure 3:   Probability (P) and frequency (F) distribution function of runoff from growing medium varies the maximum water content (w_{max}).

Expressing the probability of vegetation survival in terms of average return intervals, ARI = 1/(1-P), it varies from 3 years with z_g = 50 mm, to 4 years with z_g = 200 mm (Table 3); also increasing till five times the growing medium thickness, from 100 mm to 500 mm ARI did not improve. To achieve higher ARIs, an additional water supply to the green roof, by way of an irrigation system, is needed.

Fig. 3 shows the runoff PDF varying green roof maximum retention capacity.

Assuming that the maximum retention capacity by vegetation is of a few millimetres, the maximum retention capacity of the drainage layer is equal to 50 mm, and that with the growing medium thickness z_g = 100 mm, the maximum retention capacity of the layer was 58 mm, the maximum retention capacity resulted equal to about 110 mm; for this value the probability of runoff is very low, and it can be considered optimum to guarantee a good performance of the green roof, both in terms of stormwater control and fulfilment of water demand by the non-irrigated vegetation.

5 CONCLUSIONS

The proposed method allows one to estimate the probability of runoff from green roofs and the probability of survival of the vegetation cover without irrigation. The equations developed enable designers to link these probabilities with the thickness of growing medium, considering both the vegetation type and the climate features of the site.

An important improvement of the proposed method is that it allows us to consider the effects of chained rainfall events in the evaluation of the probabilities, without the need for continuous simulation of the hydrological processes. This makes the results more realistic and reliable, and the application easier and cheaper, in terms of time and data needs. The developed equations can be a valid aid for green roof design, as they allow us to define the thickness of the growing medium for different levels of risk.

The example of application to a case study in Milan, Italy showed a good fit of the results obtained by the proposed formulas and continuous simulation of observed data. An interesting result was that for both the probability of runoff and the probability of having minimum water content in the growing medium for vegetation to survive without irrigation, increasing growing medium thickness beyond a certain threshold does not provide significant advantages, in terms of green roof performance, but only economical disadvantages. Upon application to our case study, the optimal value was around 100 mm. Although the probabilistic model is of general value and applicability, numerical results were related and limited to the climatic features of the case study. Future developments of the proposed model will then also consider the application to other case studies, in different climatic contexts.

REFERENCES

[1]   Lee, E.H., Choi, Y.H. & Kim, J.H., Real-time integrated operation for urban streams with centralized and decentralized reservoirs to improve system resilience. *Water*, **11**(1), p. 69, 2019. https://doi.org/10.3390/w11010069.

[2]   Carter, T.L. & Rasmussen, T.C., Hydrologic behaviour of vegetated roofs. *Journal of the American Water Resources Association*, **42**(5), pp. 1261–1274, 2006.

[3]   Getter, K.L., Rowe, D.B. & Andresen, J.A., Quantifying the effect of slope on extensive green roof stormwater retention. *Ecological Engineering*, **31**(4), pp. 225–231, 2007.

[4]   Li, Y. & Babcock Jr., R.W., Green roof hydrologic performance and modelling: A review. *Water Science and Technology*, **69**(4), pp. 727–738, 2014.

[5]   Li, Y. & Babcock Jr., R.W., Modelling hydrologic performance of a green roof system with HYDRUS-2D. *Journal of Environmental Engineering*, **141**(11), 04015036, 2015.

[6] Palla, A., Gnecco, I. & Lanza, L., Compared performance of a conceptual and a mechanistic hydrologic model of a green roof. *Hydrological Processes*, **26**(1), pp. 73–84, 2012.

[7] Locatelli, L., Mark, O., Mikkelsen, P.S., Arnbjerg-Nielsen, K., Bergen, J.M. & Binning, P.J., Modelling of green roof hydrological performance for urban drainage applications. *Journal of Hydrology*, **519**, pp. 3237–3248, 2014.

[8] Stovin, V., Vesuviano, G. & De-Ville, S., Defining green roof detention performance. *Urban Water Journal*, **14**(6), pp. 574–588, 2017.

[9] Palermo, S.A., Turco, M., Principato, S. & Piro, P., Hydrological effectiveness of an extensive green roof in Mediterranean climate. *Water*, **11**(7), p. 1378, 2019.

[10] Piro, P., Carbone, M., De Simone, M., Maiolo, M., Bevilacqua, P. & Arcuri, N., Energy and hydraulic performance of a vegetated roof in sub-Mediterranean climate. *Sustainability*, **10**(10), p. 3473, 2018. https://doi.org/10.3390/su1010347.

[11] Herrera, J., Flamant, G., Gironás, J., Vera, S., Bonilla, C.A., Bustamante, W. & Suárez, F., Using a hydrological model to simulate the performance and estimate the runoff coefficient of green roofs in semiarid climates. *Water*, **10**(2), p. 198, 2018.

[12] Peng, L.L. & Jim, C.Y., Seasonal and diurnal thermal performance of a subtropical extensive green roof: The impacts of background weather parameters. *Sustainability*, **7**(8), pp. 11098–11113, 2015. https://doi.org/10.3390/su70811098.

[13] Hakimdavar, R., Culligan, P.J., Finazzi, M., Barontini, S. & Ranzi, R., Scale dynamics of extensive green roofs: Quantifying the effect of drainage area and rainfall characteristics on observed and modeled green roof hydrologic performance. *Ecological Engineering*, **73**, pp. 494–508, 2014. https://doi.org/10.1016/j.ecoleng.2014.09.080.

[14] Nawaz, R., McDonald, A. & Postoyko, S., Hydrological performance of a full-scale extensive green roof located in a temperate climate. *Ecological Engineering*, **82**, pp. 66–80, 2015. https://doi.org/10.1016/j.ecoleng.2014.11.061.

[15] Zhang, S. & Guo, Y., Analytical probabilistic model for evaluating the hydrologic performance of green roofs. *Journal of Hydrological Engineering*, **18**(1), pp. 19–28, 2013. https://doi.org/10.1061/(ASCE)HE.1943-5584.0000593.

[16] Berndtsson, J.C., Green roof performance towards management of runoff water quantity and quality: A review. *Ecological Engineering*, **36**, pp. 351–360, 2010. https://doi.org/10.1016/j.ecoleng.2009.12.014.

[17] Guo, Y. & Adams, B.J., Hydrologic analysis of urban catchments with event-based probabilistic models. Runoff volume. *Water Resources Research*, **34**(12), pp. 3421–3431, 1998. https://doi.org/10.1029/98WR02449.

[18] Guo, Y. & Adams, B.J., Hydrologic analysis of urban catchments with event-based probabilistic models. Peak discharge rate. *Water Resources Research*, **34**(12), pp. 3433–3443, 1998. https://doi.org/10.1029/98WR02448.

[19] Guo, Y. & Adams, B.J., An analytical probabilistic approach to sizing flood control detention facilities. *Water Resources Research*, **35**(8), pp. 2457–2468, 1999. https://doi.org/10.1029/1999WR900125.

[20] Adams, B.J. & Papa, F., *Urban Stormwater Management Planning with Analytical Probabilistic Models*, John Wiley: New York, 2000.

[21] Raimondi, A. & Becciu, G., An analytical probabilistic approach to size cisterns and storage units in green buildings. *Proceedings of CCWI2011*, 2011.

[22] Raimondi, A. & Becciu, G., Probabilistic design of multi-use rainwater tanks. *Procedia Engineering*, **70**, pp. 1391–1400, 2014. https://doi.org/10.1016/j.proeng.2014.02.154.

[23] Raimondi, A. & Becciu, G., Probabilistic modeling of rainwater tanks. *Procedia Engineering,* **89**, pp. 1493–1499, 2014. https://doi.org/10.1016/j.proeng.2014.11.437.

[24] Becciu, G., Raimondi, A. & Dresti, C., Semi-probabilistic design of rainwater tanks: A case study in Northern Italy. *Urban Water Journal,* **15**(3), pp. 192–199, 2018. https://doi.org/10.1080/1573062X.2016.1148177.

[25] Guo, Y. & Gao, T., Analytical equations for estimating the total runoff reduction efficiency of infiltration trenches. *Journal of Sustainable Water Built Environment,* **2**(3), 2016. https://doi.org/10.1061/JSWBAY.0000809.

[26] Wang, J. & Guo, Y., Proper sizing of infiltration trenches using closed-form analytical equations. *Water Resources Management,* **34**, pp. 3809–3821, 2020.

[27] Zhang, S. & Guo, Y., An analytical equation for evaluating the stormwater volume control performance of permeable pavement systems. *Journal of Irrigation Drainage Engineering,* **141**(4), 2014. https://doi.org/10.1061/(ASCE)IR.1943-4774.0000810.

[28] Zhang, S. & Guo, Y., Stormwater capture efficiency of bioretention systems. *Water Resources Management,* **28**(1), pp. 149–168, 2014. https://doi.org/10.1007/s11269-013-0477-y.

[29] Guo, Y. & Gao, T., Analytical equations for estimating the total runoff reduction efficiency of infiltration trenches. *Journal of Sustainable Water Built Environment,* **2**(3), 2016. https://doi.org/10.1061/JSWBAY.0000809.

[30] Guo, Y. & Zhang, S., Runoff reduction capabilities and irrigation requirements of green roofs. *Water Resource Management,* **28**, pp. 1363–1378, 2014. https://doi.org/10.1007/s11269-014-0555-9.

[31] Zhang, S. & Guo, Y., Analytical probabilistic model for evaluating the hydrologic performance of green roofs. *Journal of Hydrological Engineering,* **18**(1), pp. 19–28, 2013. https://doi.org/10.1061/(ASCE)HE.1943-5584.0000593.

[32] Raimondi, A. & Becciu, G., On pre-filling probability of flood control detention facilities. *Urban Water Journal,* **12**(4), pp. 344–351, 2015. https://doi.org/10.1080/1573062X.2014.901398.

[33] Raimondi, A. & Becciu, G., On the efficiency of stormwater detention tanks in pollutant removal. *International Journal of Sustainable Development Planning,* **12**, pp. 44–154, 2017. https://doi.org/10.2495/SDP-V12-N1-144-154.

[34] Becciu, G. & Raimondi, A., Factors affecting the pre-filling probability of water storage tanks. *WIT Transactions on Ecology and the Environment,* vol. 164, WIT Press: Southampton and Boston, pp. 473–484, 2012.

[35] Becciu, G. & Raimondi, A., Probabilistic analysis of the retention time in stormwater detention facilities. *Procedia Engineering,* **119**, pp. 1299–1307, 2015.

[36] Becciu, G. & Raimondi, A., Probabilistic modeling of the efficiency of a stormwater detention facility. *International Journal of Sustainable Development Planning,* **10**(6), pp. 795–805, 2015. https://doi.org/10.2495/sdp-v10-n6-795-805.

[37] Becciu, G. & Raimondi, A., Probabilistic analysis of spills from stormwater detention facilities. *WIT Transaction of the Built Environment,* vol. 139, WIT Press: Southampton and Boston, pp. 159–170, 2014.

[38] United States Environmental Protection Agency (EPA), Methodology for analysis of detention basins for control of urban runoff quality. U.S. EPA Office of Water, Nonpoint Source Branch report, p. 51, 1986.

[39] Adams, B.J., Fraser, H.G., Charles, D., Howard, D. & Hanafy, M.S., Meteorological data analysis for drainage system design. *Journal of Environmental Engineering,* **112**(5), pp. 827–848, 1986.

[40] Eagleson, P.S., Climate, soil and vegetation: The distribution of annual precipitation derived from observed storm sequences. *Water Resources Research*, **14**(5), pp. 713–721, 1978.

[41] Bedient, P.B., Huber, W.C. & Vieux, B.E., *Hydrology and Floodplain Analysis*, 2008.

[42] Bacchi, B., Balistrocchi, M. & Grossi, G., Proposal of a semi-probabilistic approach for storage facility design. *Urban Water Journal*, **5**(3), pp. 195–208, 2008.

[43] Palermo, S.A., Turco, M., Principato, S. & Piro, P., Hydrological effectiveness of an extensive green roof in Mediterranean climate. *Water*, **11**(7), p. 1378, 2019.

[44] Wolf, D. & Lundholm, J.T., Water uptake in green roof microcosms: Effects of plant species and water availability. *Ecological Engineering*, **33**(2), pp. 179–186, 2008.

[45] Kaiser, D., Kohler, M., Schmidt, M. & Wolff, F., Increasing evapotranspiration on extensive green roofs by changing substrate depths, construction and additional irrigation. *Buildings*, **9**(7), p. 173, 2009.

"WHAT'S THE CARBON CONTENT OF YOUR FOOD?": DEVELOPMENT OF AN INTERACTIVE ONLINE *FOODPRINT* SIMULATOR

KHADIJA BENIS[1], WALEED ALHAYAZA[2], ADNAN ALSAATI[2] & CHRISTOPH REINHART[1]
[1]Department of Architecture, Massachusetts Institute of Technology, USA
[2]King Abdulaziz City for Science and Technology, Saudi Arabia

ABSTRACT
As part of their sustainability agendas, cities are promoting urban food production as a means to reduce the demand for agricultural land elsewhere and shorten food miles. However, from an environmental sustainability standpoint, some assessments have shown that increasing self-sufficiency through local production for certain crops is not necessarily more sustainable than the current practice of importing them. Therefore, in addition to quantifying the potential for food self-sufficiency of cities through urban agriculture, there is a need to assess whether such scenarios are indeed more resource efficient than existing supply chains. For that purpose, a fundamental step in the sustainability assessment of alternative local food supply practices is the assessment of existing supply chains, to be used as baseline scenarios of the analysis. In our previous work, we referred to these baseline scenarios as "Urban Foodprints" (UF), a concept that expresses resource consumption and environmental impacts associated with the urban food system, from agricultural production to distribution and consumption. The very first step in building the UF of a given city is collecting information on its actual food demand, which is oftentimes a challenging task due to the scarcity of reliable data sources on food consumption. To fill this gap, this paper presents the development of an online food intake survey that collects a detailed snapshot of the self-reported dietary habits of respondents and probes to what extent their consumption choices are guided by sustainability concerns. By providing respondents with an individualized carbon content analysis of their food intake upon survey completion, this study further aims to raise awareness on the environmental impacts of our food choices as well as which food choices are most impactful.
Keywords: survey, food consumption, food system, environmental impacts, urban foodprint.

1 INTRODUCTION
According to the Intergovernmental Panel on Climate Change (IPCC), the food system as a whole – growing, harvesting, processing, packaging, transporting, marketing, consumption and disposal of food and food-related items – contributes to approximately 26% of global anthropogenic greenhouse gas (GHG) emissions. This number not only accounts for land use change (i.e., deforestation and peatland degradation), but also for "outside farm gate" emissions from energy, transport and industry sectors for food production (see Fig. 1). In fact, the food system's share of anthropogenic GHG emissions is nearly twice that of the entire transportation sector (14%) including air travel. With raising societal awareness related to sustainable living, the environmental footprint of the food sector has been under increasing scrutiny by the scientific community over the past decade. The method of choice to study the underlying processes is life cycle assessment [1]. The major portion of emissions (82%) stems from agricultural production (i.e., emissions of nitrous oxide resulting from the use of fertilizers; emissions of methane from livestock) and from the loss of carbon sinks as the expansion of agricultural land for crop cultivation and grazing of livestock is propelling deforestation. Outside the farm gate, pre- and post-production sectors represent 18% of food system emissions. All these numbers show how mitigating emissions from the global food system could have a major impact in the fight against climate change. However, trends in the sector are currently moving in the opposite direction and GHG emissions are actually

WIT Transactions on Ecology and the Environment, Vol 243, © 2020 WIT Press
www.witpress.com, ISSN 1743-3541 (on-line)
doi:10.2495/UA200111

projected to increase, driven by population and income growth and changes in consumption patterns [2].

As part of their sustainability agendas, some cities are promoting urban food production as a means to reduce the demand for agricultural land elsewhere and shorten food miles [4]. However, from an environmental sustainability standpoint, some assessments have shown that increasing self-sufficiency through local production for some crops is not always more sustainable than current practice. For example, a study in the UK showed that producing greenhouse strawberries in London may have a higher carbon footprint than importing Spanish greenhouse strawberries [5]. Therefore, in addition to quantifying the potential for food self-sufficiency of cities through urban cultivation, there is a crucial need to assess the extent to which such scenarios are more resource efficient than existing supply chains. For that purpose, a fundamental step in the sustainability assessment of alternative local food supply practices is the assessment of existing supply chains, to be used as baseline scenarios of the analysis. In our previous work, we referred to these baseline scenarios as "Urban Foodprints" (UF), a concept that expresses resource consumption and environmental impacts associated with the urban food system, from agricultural production to distribution and consumption [6]. The very first step in building the Urban Foodprint of a given city is collecting information on its actual food demand, which is oftentimes a challenging task due to the scarcity of reliable data sources on food consumption.

In our previous studies, we have been using the Food Balance Sheets (FBS) of the United Nations' Food and Agriculture Organization (FAO) as a proxy for food consumption. Online FBS datasets are gathered by FAO every year for 185 nations for about 100 food commodity groups that are supplied for human consumption. Gross national food supply in a given reference period of a country is calculated from the total quantity of food produced plus the total quantity imported, adjusted for changes at national food stock levels and exports. Net food availability is calculated by subtracting the amounts used for animal feed, seeds, industrial or other purposes and losses in the supply chain (see eqn (1) below). This net value is then divided by the country's population estimate to obtain a final figure describing the availability of food commodities, expressed as kilograms per capita per year. This per capita information can also be linked to food composition data and presented as per capita energy intake (kilocalories per day), protein intake (grams per day) and fat intake (grams per day) [7].

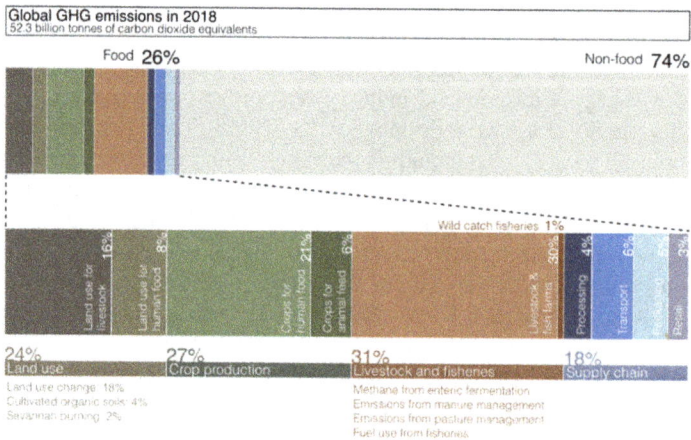

Figure 1: Global GHG emissions versus food system GHG emissions [3].

$$NFS_{COUNTRY} = S_{ST} + (F_D + F_I) - (F_E + S + AF + W + O_{NFU}) - S_{END}, \qquad (1)$$

where: $NFS_{COUNTRY}$ is the net national food supply in a given reference period of a country; S_{ST} are the starting food stocks; F_D is the quantity of food produced domestically; F_I is the quantity of food imported; F_E is the quantity of food exported; S is seed; AF is animal feed; W is waste (at farm gate); O_{NFU} are other non-food uses; S_{END} are the ending stocks.

The FAO FBS are generally utilized in research to screen worldwide dietary habits, trends in national food availability, and the capability of a country's supply to meet nutritional needs. However, using these datasets as a proxy for food consumption has some limitations, not only because estimates are based on country statistics which themselves may be subject to methodological errors, but also because domestic and retail food waste are not accounted for. Moving forward, in order to improve the accuracy of our data, this manuscript presents results from an online survey on dietary habits of participants. It is organized as follows: Section 2 synthesizes the existing methods for surveying food intakes and identifies research gaps. Section 3 describes the methodology underlying our survey, followed by the discussion of preliminary results in section 4, and expected outcomes in Section 5.

2 METHODS FOR SURVEYING INDIVIDUAL FOOD INTAKE

Conventional approaches for direct data collection of dietary patterns of individuals can be divided into two categories, retrospective and prospective tools. While the former measures past food intake via self- or interviewer-administered recall questionnaires, the latter records food intake at the time of consumption. Table 1 organizes the existing methods following this distinction reviewing in turn both methods' respective descriptions, administration modes, time frames, and limitations as acknowledged in the literature [7].

Bias and misreporting are common issues to all these approaches, at varying degrees. For instance, retrospective methods heavily rely on the respondents' memory and can therefore not only lead to errors when reporting on frequency and portion size estimation, but respondents may also be selective with the foods they choose to report during the recall. On the other hand, prospective methods are burdensome to respondents as they involve documenting (and sometimes weighing) every food and beverage consumed at each meal during a predefined period of time, and therefore may either interfere with respondents' normal eating habits or lead to a decreasing reliability of records over time as food intakes can be missed or less accurately recorded. Furthermore, as both recall- and record-based approaches require multiple days of data collection to adequately represent the habitual food intake of individuals, all these conventional approaches tend to be time-intensive for both respondents and researchers. Consequently, for the latter, all of these methods are labour-intensive (and therefore cost-intensive), either requiring well-trained interviewers or involving great amounts of time spent on data entry and food matching with food composition data.

Faced with all these limitations of conventional methods, researchers in the food and nutrition field are more and more looking into integrating innovative technologies to improve dietary assessment (see Table 2) [7].

These new methods seek to replace pen and paper approaches by relying on the use of technologies such as mobile phones or interactive computer software, to reduce memory bias and gather more reliable information. Compared with conventional dietary assessment methods, such approaches reduce the cost of data collection and processing, as they allow researchers to automatically process the collected information and get real-time feedback.

Table 1: Conventional food intake survey methods.

	1.1. Retrospective methods			1.2. Prospective methods		
Method	**1.1.1.** Dietary history	**1.1.2.** Food Frequency Questionnaire (FFQ)	**1.1.3.** 24-hour recall	**1.2.1.** Estimated food records	**1.2.2.** Weighed food records	**1.2.3.** Duplicate meal method
Description	Describe annual food intake and its variation over a long period of time (6 months–1 year): (1) interview; (2) food list; (3) 3-day record with portion size estimates as cross-checks	Assess the frequency with which foods and/or food groups are eaten over a certain period. Questionnaire includes a food list and a frequency category section	Respondents are asked, by a trained nutritionist or dietitian, to recall and report all food and beverages consumed over the past 24 hours	Respondents are instructed to document all food and beverages consumed during a predefined period (e.g., 1 to 7 days)	Same method as estimated food records, + respondents are asked to weigh using weighing scales	Respondents set aside duplicate portions of all foods consumed, weighed and sent to a laboratory for chemical analysis to determine nutrient content
Administration	Self- or interviewer-administered	Self- or interviewer-administered	Interviewer-administered	Self-administered	Self-administered	Self-administered
Time frame	Multi-day recall	Daily, weekly or monthly	Multi-day recall	Multi-day record	Multi-day record	Multi-day record
Relies on respondents' memory	Yes	Yes	Yes			
May interfere with normal eating habits				Yes	Yes	Yes
Time-intensive (respondents)	+	+	+	++	++	+++
Time-intensive (researchers)	++	++	+++	++	++	+++
Costly for respondents						+++
Costly for researchers	++	++	+++	++	++	+++

"+" is lowly; "++" is moderately; and "+++" is highly.

Table 2: Innovative food intake survey methods.

	2. Innovative methods				
Method	**2.1.** Personal Digital Assistant (PDA)	**2.2.** Image-assisted methods (i.e., digital photographs)	**2.3.** Image-assisted methods (i.e., mobile-based technologies)	**2.4.** Interactive computer and web-based technologies	**2.5.** Scan- and sensor-based technologies
Description	Specially-designed dietary software program used to register dietary intake. Participants receive training on how to use the device and record their food intake by selecting food items from a list	Method that uses images (i.e. photographs, videos) of food collected during eating episodes to enhance accuracy and reduce respondent burden and memory bias	Allow users to shoot images or make voice records using a smartphone or tablet	Involve the use of interactive dietary assessment programs installed on a computer; the "web-based" moniker refers to these devices being connected to the internet	Participants scan the barcodes of purchased food items; sensor-based are designed to be memory-independent and almost passive for participants

3 THE FOODPRINT SIMULATOR – MATERIALS AND METHODS

Whereas conventional pen and paper methods (Table 1) are usually burdensome and costly to both users and researchers (and potentially less accurate), technology-based approaches (Table 2) are increasingly being used to monitor food intake. However, even within this group of approaches, choosing a survey method depends on the objective of the particular study that is being undertaken. This section describes the goals and methodology underlying the development of the *FOODprint Simulator*, as well as the data sources used for that purpose.

3.1 Objective

As mentioned in Section 1, the goals of this study are two-fold: (1) being able to define accurate baseline scenarios of food consumption based on actual feedback from consumers; and (2) raising awareness on the environmental impacts of food. For this purpose, our research requirements not only imply reaching a wide audience in a cost-effective manner, but also ensuring an adequate level of accuracy. We therefore chose to develop the *FOODprint Simulator* within the methodological framework of the "Innovative methods" group (see Fig. 2), as a survey that consists of a web-based, self-administered questionnaire, with a core part inviting respondents to build typical meals that represent their daily food consumption habits. Compared to other personal food trackers available on the market, the novelty of this tool lies in providing an additional set of information to the users besides nutritional data, i.e., the carbon content of their diets.

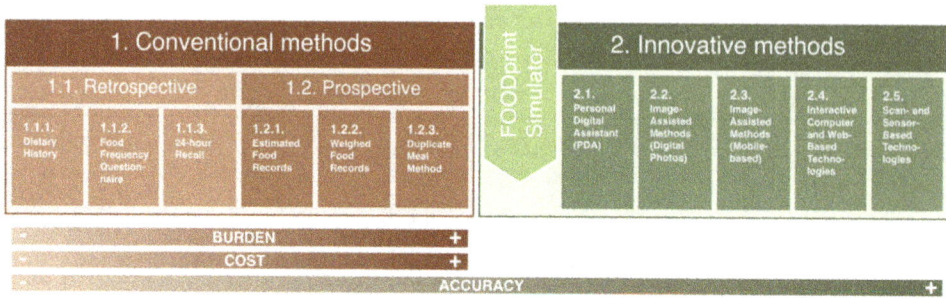

Figure 2: Methodological framework of the *FOODprint Simulator*.

3.2 Questionnaire structure

The *FOODprint Simulator* questionnaire contains three Sections, followed by a final page displaying the results. Section 1, entitled "About you," collects general demographic information, such as gender, age, place of residence, level of education, and general activity level. From gender, age and activity level data, a "normal" daily calorie intake is generated to help participants gauge whether their entries are plausible. Section 2, entitled "What's on your plate?" is the core of the survey, where respondents are invited to build as many typical meals and they wish, in a way that consistently represents their food intake. To do so, they can select food and beverage items from a list (that also provides pictures of serving sizes for reference) and drop them into a food tray. To adjust portion sizes of food, they can add more servings; drink sizes can be adjusted as well. Throughout the whole process of composing meals, respondents are able to check their average calorie count as well as their carbon emissions. Section 3, entitled "What are your shopping preferences", gathers some additional information on the respondents' grocery shopping habits, containing questions about the preferences that guide their choice of products when purchasing food. This section will inform us on the potential links of sustainability concerns with food consumption habits, and the willingness of consumers to pay a premium for sustainably sourced products. Finally, upon completion of the survey, respondents get to visualize the total carbon content of their food, as well as its breakdown into the main six food groups of the FAO classification. To better understand their results, they can compare them to the average carbon footprint of food supply in their country and worldwide (see Fig. 3).

3.3 Underlying calculations and datasets

The conception of Section 2, "What's on your plate?," involved the development of an underlying database and calculation system that provides instantaneous feedback to the user, as he is defining his meals. Efforts were made – on the whole survey in general and on this section in particular – to build an attractive and intuitive interface that would arouse the respondents' curiosity and make them want to dedicate some time and attention to building typical meals and exploring the associated nutritional and environmental information provided instantaneously by the tool. Behind this interface, our database uses the FAO taxonomy, which classifies food into six categories (meat; dairy and eggs; grain; produce; sugar and fat; and other), subdivided into 20 sub-groups (see Fig. 4).

Each time the user drops a food or beverage item into his tray, carbon and calorie contents are calculated within the database, according to the composition of the food item and using reference values of calorie content and carbon intensity, multiplied by the selected portion size. Average carbon intensities of all food items were sourced from various Life Cycle Assessment (LCA) studies.

Additionally, to enable the user to compare his final FOODprint to the average carbon footprint of food supply in his country and worldwide, the database contains country-specific data, calculated using food and calorie supply data from the FAOSTAT database.

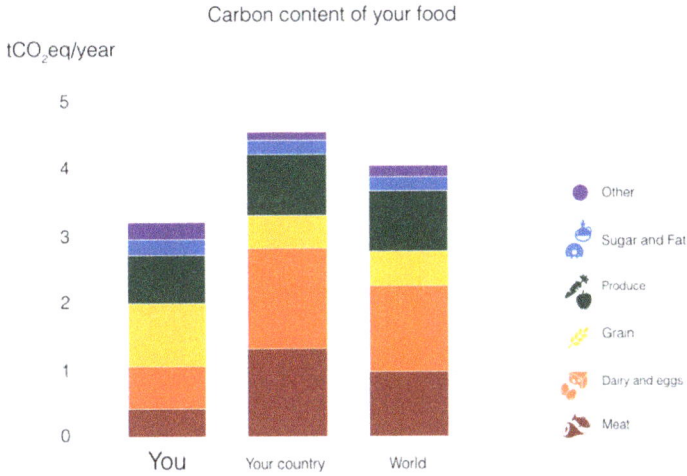

Figure 3: Final results displayed at the end of the survey.

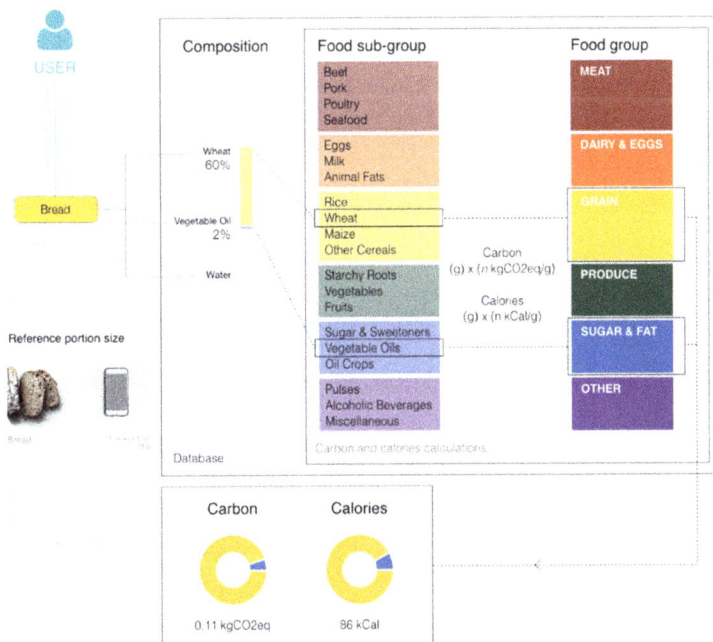

Figure 4: Workflow for an example food item.

4 APPLICATION OF THE SURVEY AND PRELIMINARY RESULTS

The *FOODprint Simulator* was published online in March 2020. Its cover page provided a short description of the objective of the study. The confidentiality of replies was highlighted, and an average completion time was suggested. The survey was disseminated through several social media platforms. It was completed by 262 respondents between March and May. This section presents some preliminary results from this first deployment phase.

4.1 Distribution strategy

As of 2019, 59% of the world's population uses the internet, representing a large opportunity for the emerging online data collection method. In comparison to conventional survey techniques, online surveys are a cost- and time-efficient way of collecting data, allowing researchers to get feedback faster and from a broader audience. At the same time, data processing and analysis can be done in real-time as responses are being collected, and anonymity can encourage respondents to provide more honest answers. Furthermore, 80% of internet users are also active social media users, hence our distribution strategy, not only via emails through our mailing lists, but also through the main social media platforms (e.g., Twitter, Instagram, Facebook).

4.2 Preliminary results

4.2.1 Demographic characterization

The respondents were from 17 different countries, with a predominance of residents of Saudi Arabia (54%) and the United States (32%) – the two pilot countries where most of our dissemination efforts were focused. Young and middle-aged were the prevalent age groups, with 31% of the respondents belonging to the 25–34 group; 30% to the 35–49 group; and 26% to the 19–24 group. Almost two-thirds of the respondents were female (63%); 88% of the respondents have a higher education degree. Finally, most of the respondents reported low to moderate activity levels – 44% exercised once a week or less; 42% exercised 1 to 3 times a week; and only 14% exercised at least 5 times a week.

4.2.2 Carbon FOODprint

The average carbon content of the respondents' diets, by age group, are shown in Fig. 5, with overall FOODprints ranging between 2.1 and 2.9 tCO_2eq/year. When broken down by food items, dairy, meat and eggs dominate across all age groups, accounting for 54 to 66% of carbon FOODprints.

At the same time, meat-related carbon emissions decreased among older populations, accompanied by a slight increase in produce-related emissions. Such a trend might reflect healthier dietary habits related to ageing.

4.2.3 Sustainability awareness

The third section of the survey allows decision-makers to gauge the potential links of sustainability concerns with food consumption habits, and the willingness of consumers to pay a premium for sustainably sourced products. Across all age groups, respondents showed a preference for local, organic, and low-carbon foods, although some of them changed their answers when asked about their willingness to pay a 50% premium for these products (see Fig. 6).

tCO$_2$eq/year

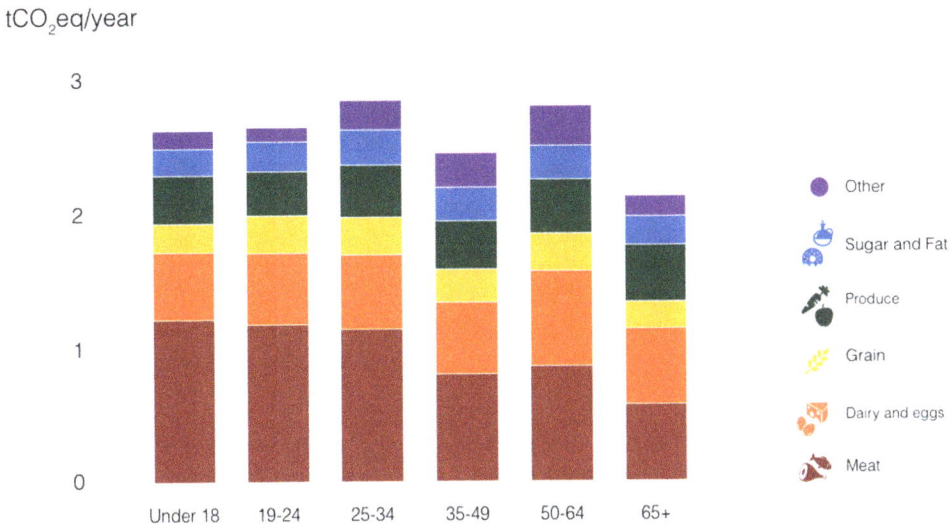

Figure 5: Carbon content of respondents' diets, by age group.

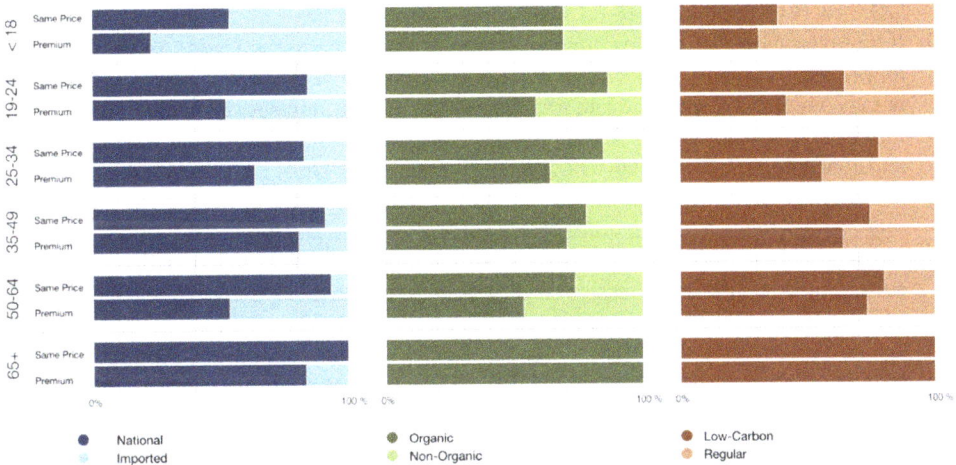

Figure 6: Shopping preferences, by age group.

5 NEXT STEPS

Carbon accounting is largely used, not only as a sustainability assessment tool of products and companies, but also as a powerful instrument to raise awareness about climate change. Given that food-related carbon emissions represent one quarter of total anthropogenic emissions, it is crucial to estimate accurately these emissions in order to define paths of mitigation. In that sense, the *FOODprint Simulator* is a powerful tool both for decision-makers and for sustainability consultants, willing to quantitatively assess the impacts of local food production in a given neighbourhood or city, against current scenarios.

Recently, during a major event of the Expo 2015, the city of Milan promoted an international protocol, engaging the largest number of world cities for the development of food systems, based on the principles of sustainability. The Milan Urban Food Policy Pact was signed by more than 200 cities all over the world. The simulator can help collecting useful information on dietary patterns and food sustainability awareness, for policy makers to understand better the demand of their residents and be able to tailor policies accordingly.

The preliminary results presented here were aimed at troubleshooting the survey and improving it through the feedback obtained from the first set of respondents. We will further work together with institutional, academic and corporate study groups, where the survey will be administered in a more focused manner, in sessions where a trained person or educator will guide respondents through the process of defining their food intake. Through all these channels, the survey will be deployed and available for at least one full year

ACKNOWLEDGEMENTS
Generous support for this work has been provided by the King Abdulaziz City for Science and Technology (KACST) as part of the research under the Center for Complex Engineering Systems (CCES) at MIT and KACST through the Urban Water, Energy and Food project.

REFERENCES
[1] Benis, K. & Ferrão, P., Potential mitigation of the environmental impacts of food systems through urban and peri-urban agriculture (UPA): A life cycle assessment approach. *Journal of Cleaner Production*, **140**, pp. 784–795, 2017. DOI: 10.1016/j.jclepro.2016.05.176.
[2] Intergovernmental Panel on Climate Change, Climate change and land: An IPCC special report on climate change, desertification, land degradation, sustainable land management, food security, and greenhouse gas fluxes in terrestrial ecosystems, 2019.
[3] Poore, J. & Nemecek, T., Reducing food's environmental impacts through producers and consumers. *Science*, **360**(6392), pp. 987–992, 2018. DOI: 10.1126/science.aaq0216.
[4] Baker, L. & de Zeeuw, H., Urban food policies and programmes: An overview. *Cities and Agriculture: Developing Resilient Urban Food Systems*, pp. 26–55, 2015.
[5] Kulak, M., Graves, A. & Chatterton, J., Reducing greenhouse gas emissions with urban agriculture: A life cycle assessment perspective. *Landscape and Urban Planning*, **111**(1), pp. 68–78, 2013. DOI: 10.1016/j.landurbplan.2012.11.007.
[6] Benis, K., Gashgari, R., Alsaati, A. & Reinhart, C., Urban foodprints (UF): Establishing baseline scenarios for the sustainability assessment of high-yield urban agriculture. *International Journal of Design & Nature and Ecodynamics*, **13**(4), pp. 349–360, 2018. DOI: 10.2495/DNE-V0-N0-1-12.
[7] FAO, Dietary assessment: A resource guide to method selection and application in low resource settings, Rome, 2018.

MEASUREMENT OF TECHNICAL EFFICIENCY: A CASE STUDY OF DAILOAN-MANGO IN VIETNAM

TRUONG HONG VO TUAN KIET[1,2,3], NGUYEN THI KIM THOA[1] & PHAM THI NGUYEN[4]
[1]Mekong Delta Development Research Institute, Can Tho University, Vietnam
[2]Southeast Asian Regional Center for Graduate Study and Research in Agriculture, Philippines
[3]Institute of Agricultural and Food Policy Studies, Putra University, Malaysia
[4]School of Economics, Can Tho University, Vietnam

ABSTRACT

The findings of this study present the results of three producing seasons of DaiLoan-mango growers per year. The results showed that season 3 was the highest technical efficiency (73.10%), the second place was season 1 (62.65%), and then season 2 (58.41%). Moreover, the independent variables that explained the technical efficiency of season 1 were the pesticide, root fertiliser, leaf fertiliser and labour; season 2 were the root fertiliser, leaf fertiliser and labour; and in season 3 were the pesticide, root fertiliser, leaf fertiliser, and labour. Besides this, the important determinants of socio-economic variables that impacted positively on technical efficiency were the land area in all three seasons, market access in season 2, and credit access in season 3. However, the constraints to DaiLoan-mango production were the farming experience in season 1, and the age in season 2. Based on these findings, policy makers should focus on effective input models that would boost technical efficiency through conducting regular workshops and orchard demonstrations on using input materials effectively for mango farmers. Moreover, farmers should be empowered in land area acquisition to apply advanced technology in large-scale production more effectively.

Keywords: DaiLoan-mango, technical efficiency, stochastic frontier.

1 INTRODUCTION

The king of the fruits, mango fruit, is one of the most popular fruits over the world, scientific name of mango is Mangifera Indica. Mangoes have traditionally been cultivated both in the sub-tropics and tropics. Vietnam provided mango volume approximately 836,000 tons in 2017 [1]. It ranked fourth in terms of mango volume in Southeast Asia after Thailand, Indonesia and Philippines and was top 15 the largest mango producers in the world. In Vietnam, mango is planted in most of all provinces, especially the southern Vietnam accounts for 75% of the mango production volume and for 72% of the mango production area [2].

The findings of Loc et al. [3] showed that gross income of mango farmers is an average of 186 million VND/ household/year, in which net income is 105.4 million VND, and average household cultivation area of 0.68 ha. Most of mango growers are small-scale; therefore, they confront with difficulties such as market signals relating to demand, varieties, quality and food safety, poor technical skills. Besides, mango producers must confront with uncertainly selling price, depending on collectors. In Vietnam, there are four prevalent mango varieties such as HoaLoc-mango, Chu-mango, Xiem-mango, and DaiLoan-mango, in which DaiLoan is considered new mango varieties, has been planted for 20 years, and become popular in 10 recent years. If HoaLoc, Chu, and Xiem mangoes were grown in rural areas, DaiLoan mango would be planted not only in remote places but also developed in urban region. One of main reasons is that DaiLoan-mango has adapted well various climate and soil conditions. Thus, households in cities choose DaiLoan-mango to grow as a kind of fruit and vegetables to provide fiber, sugar, protein, vitamin, calcium for their family (it can be eaten both raw and ripe fruit form). These are all extremely important nutrients for citizen's health. This type of fruit is not only delicious but also wonderful with delicious taste. That is reason why

WIT Transactions on Ecology and the Environment, Vol 243, © 2020 WIT Press
www.witpress.com, ISSN 1743-3541 (on-line)
doi:10.2495/UA200121

DaiLoan-mango is only mango varieties to become the most well-known fruit of households in urban areas in Vietnam. Particularly, it is suitable to produce small-scale with households level to apply smart farming system via hi-tech in context of agricultural urban development more and more to contribute diversity of nutritious supply sources to citizens towards food security at household level.

Hence, the objective of the paper was to identify determinants of technical efficiency in order to alleviate deferent constraints in the mango production as well as found out effective disparities in three mango production seasons. It helps both farmers in rural area and household in urban areas improve their farming procedure in order to obtain more effectively in production.

2 MATERIALS AND METHODS

2.1 Sampling techniques

Firstly, both south-eastern region and the Mekong Delta were chosen because these are the two biggest mango production in Vietnam, as it accounts for 75% of the mango production volume and for 72% of the mango production area in Vietnam. Secondly, Dong Nai province was chosen since it occupies approximately 55% of the mango production volume and 54% of the mango production area in south-eastern region. Thirdly, Dong Thap, An Giang, Tien Giang, Hau Giang, Vinh Long, and Tra Vinh provinces were selected because, combined, they make up approximately 77% of the mango production volume and 71% of the mango production area in Mekong Delta [2]. Finally, a simple random sampling technique was used to select 732 sample observations (239, 249, and 244 for seasons 1, 2, and 3, respectively). Data collection was investigated from August to October 2018.

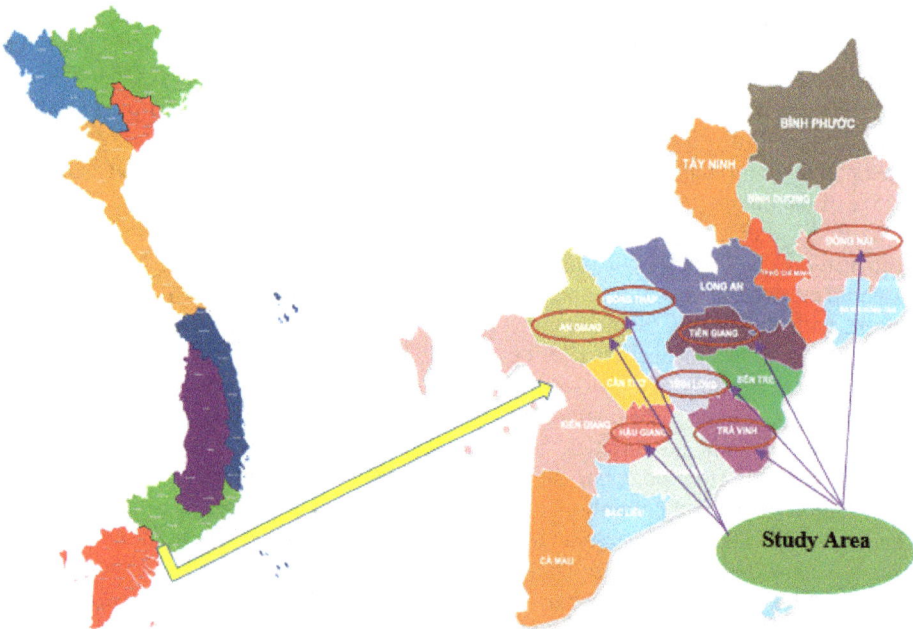

Figure 1: Study area in the southern Vietnam.

2.2 Theoretical model

Technical efficiency (TE) is the capacity of a production unit to produce a maximum level of output at the same level of inputs, or to produce a given output with minimal inputs [4], [5]. By contrast, Battese and Corra [6] claim that technical inefficiency increases when observed output is produced form given inputs to obtain less than capacity of the maximum probable. The difference in technical efficiency among farmers may be related to the manager's decision, environmental conditions (land area, rainfall, temperature and soil relative humidity), and non-technical and non-economic elements, and household characteristics that may affect a farmer's ability to use technology [7].

The measurement of farm efficiency is vital, especially for farmers in developing countries [8]. Overall, the factors that affect farmers' efficiency could be grouped into agent and structural factors. Agent factors are those linked to the farm manager, such as the educational level, family size, age, and social capital. These factors are categorised into farm-specific variables (intensity of inputs such as labour, fertilisers, and seeds), economic factors (input and output prices), and environmental factors (rain, relative humidity, and temperature).

According to Aigner et al. [9] the formulation of a stochastic frontier model in the production function could be presented as:

$$Y_i = f(X_i, \beta) \, \varepsilon_1,$$

with:

$$\varepsilon_i = V_i - U_i,$$

where:
Y_i is output of the i-th farmers;
$f(X_i, \beta)$ is an appropriate functional form;
β is a vector parameters to be estimated;
ε_i is composite error term;
V_i denotes the random error not under the control of the farmers, associated with random factors outside the farmer's control; and
U_i is the non-negative random variable associated with technical inefficiency and is identically and independently distributed as a truncated normal, with truncations at zero of the normal distribution.

The technical efficiency of the farmer is:

$$\text{Technical efficiency } (TE) = Y_i / Y_i^*,$$

where Y_i is the observed output and Y_i^* is the frontier output.

$$TE = f(X_i, \beta) \exp(V_i - U_i) / f(X_i, \beta) \exp(V_i) = \exp(-U_i),$$

this is such that $0 \leq TE_i \leq 1$.

In this case, this measure raged from zero to one, and it was used to evaluate evaluated the output of the i-th farm relative to the output of a fully efficient farm or a best practice farm using the same vector of inputs. The first step in predicting the technical efficiency according to Coelli et al. [10] is to the estimate the parameters of the stochastic production frontier. This has often taken place using the maximum likelihood estimates, which use the ordinary least squares (OLS) results as a starting point. However, the MLE are preferred over the corrected ordinary least squares (COLS) method because they have many desirable large sample (i.e. asymptotic) properties [10]. Aigner et al. [9] obtained MLE under the assumptions that

$$vi \sim (0, \sigma v^2),$$

and

$$ui \sim iidN+(0, \sigma u^2).$$

In the aforementioned equations, the v are independently and identically distributed, normal random variables with zero means and the variances σv^2 and, ui are independently and identically distributed, half-normal random variables with scale parameter $\sigma i2$. Coelli et al. [10] suggested that the finding of Battese and Corra [6] who parameterised the log-likelihood in terms of σ^2 and $\gamma = \sigma u^2/\sigma^2$, is the most appealing because if $\gamma = 0$ then all deviations from the frontier are due to statistical noise, while $\gamma = 1$ suggests that all deviations are due to technical efficiency.

2.3 Stochastic production function model

A Cobb–Douglas production function was adopted. Despite its well-known limitation, the Cobb–Douglas functional form was used. The stochastic frontier model was defined by:

$$lnYi = \beta o + \beta 1 lnX1 + \beta 2 lnX2 + \beta 3 lnX3 + \beta 4 lnX4 + \beta 5 lnX5 + Vi - Ui,$$

where:
ln = Logarithm to base e;
Yi = Mango output (kg);
βo = Constant or Intercept of the model;
$\beta 1$–$\beta 5$ = Coefficients to be estimated;
$X1$ = Pesticide quantity (litres);
$X2$ = Fungicide quantity (litres);
$X3$ = Root fertiliser quantity (kg);
$X4$ = Leaf fertiliser quantity (kg) (sprayed on mango leaves to induce flowering in mango trees);
$X5$ = Family and hired labour (man-days);
Vi = Random error term; and
Ui = Technical efficiency effect predicted by the model and the subscript i indicate the i-th farmer in the sample.

The inefficiency model based on Ogunniyi [11] was specified as

$$u_i = \alpha_0 + \sum_{r=1}^{10} \alpha_r Z_r + k,$$

where:
u_i = technical efficiency of i-th farmer;
α_0 and α_r = parameters to be estimated;
k = Truncated random variable;
Z_1 = age (year);
Z_2 = Education;
Z_3 = Farming experience (year);
Z_4 = Credit accessibility level (access =1, no access = 0);
Z_5 = Payment of agro-inputs for wholesaler (ending of crop =1, immediate payment =0);
Z_6 = Wrapping bag (wrap = 1, no wrap =0) (applied mango wrap technique against incursion of pest, insect);

Z_7 = Market accessibility level (access = 1, no access = 0);

Z_8 = Classifying sale (classification =1, no classification = 0) (selling mango is classified including: first level with best price, second level with medium price, and third level with lowest price); and

Z_9 = Land area (cong = 1,000 m^2).

The analysis of study is carried out by maximising likelihood estimation (MLE) on the STATA15.0 programme.

3 RESULTS AND DISCUSSION

3.1 Estimation procedure

The result of the MLE were presented in Table 1. The variance ratio parameter (γ) was statistically greater than zero and equal 0.6124; 0.6601 and 0.5853 for seasons 1, 2 and 3

Table 1:   Maximum likelihood estimates of the stochastic production function. *(Source: Field Survey Data, 2018.)*

Variables	Season 1		Season 2		Season 3	
	Coef.	SE	Coef.	SE	Coef.	SE
(Y): Ln yield (kg)						
Constant	5.2212***	0.3934	6.3316***	0.3717	6.7366***	0.7461
(X₁) Ln pesticide (liters)	0.1424***	0.0450	0.1579	0.0510	0.2054***	0.0508
(X₂) Ln fungicide (liters)	0.0047	0.0473	0.0943	0.0599	−0.0157	0.0502
(X₃) Ln root fertiliser (kg)	0.1328***	0.0518	−0.0515***	0.0451	0.0805**	0.0334
(X₄) Ln leaf fertiliser (kg)	0.2841***	0.0702	0.3316***	0.0601	0.1008**	0.0420
(X₅) Ln labour (man day)	0.3312***	0.0795	0.2550***	0.0945	0.1883***	0.0675
Diagnostic statistics						
Prob > χ2	0.0000		0.0000		0.0000	
Sigma square (σ²)	1.2568		1.2260		1.0432	
Lamda (λ)	1.0735		1.0441		0.5323	
Sigma_v (σᵥ)	0.6979		0.6455		0.6577	
Sigma_u (σᵤ)	0.8773		0.8996		0.7814	
Gamma (γ)	0.6124		0.6601		0.5853	
Log-likelihood function	−270.31		−327.87		−306.91	
Number of obs (N)	239		249		244	

Parameter gamma $\gamma = \sigma_u^2 / (\sigma_u^2 + \sigma_v^2)$. Sigma square $\sigma^2 = \sigma_u^2 + \sigma_v^2$.
* Significant at the 10% level, ** significant at the 5% level, *** significant at the 1% level.

respectively, thereby suggesting that 61.24%, 66.01%, and 58.53% of variation in seasons 1, 2 and 3 respectively, which caused TE of the sampled growers rather than random variability.

In the first season, the result showed that the coefficients of the pesticide, root fertiliser, leaf fertiliser and labour were positive and statistically significant at the 1% level. The positive relationship with yield proposed that a 10% increase in the pesticide, root fertiliser, leaf fertiliser and labour would result in 1.424%, 1.328%, 2.841% and 3.312% respectively, improvement in yield of DaiLoan-mango farmers.

In season 2, there were positively signed and significant in the coefficients of the leaf fertiliser, and labour at the 1% significance level, thus rising 10% of these variables would increase yield of DaiLoan-mango in 3.316%, and 2.550%, respectively. Meanwhile, the root fertiliser variable was negative effect on DaiLoan-mango yield at the 1% significance level. It meant that a 10% gain in root fertiliser quantity would decrease 0.515% of DaiLoan-mango yield.

In season 3, the results demonstrated that the coefficient explanatory variables of the pesticide, and labour in the stochastic production function were positively significant at the 1% level, and the root fertiliser, and leaf fertiliser were significant at the 5% level, thereby implying that a 10% increase in the pesticide, root fertiliser, leaf fertiliser, and labour would lead to 2.054%, 0.805%, 1.008% and 1.883% increase in mango yield for DaiLoan-mango growers, respectively.

3.2 Analysis of technical efficiency

Table 2 shows the elements influencing TE of DaiLoan-mango gardeners in Vietnam in all three seasons. The aim of estimating to identify the relationship between TE and socioeconomic characteristics.

Table 2: Estimated technical efficiency values. *(Source: Field Survey Data, 2018.)*

Variable	Season 1		Season 2		Season 3	
	Coef.	SE	Coef.	SE	Coef.	SE
Constant	0.5937***	0.0497	0.5580***	0.0500	0.6978***	0.0228
Age (Z1)	−0.0004	0.0006	−0.0011*	0.0007	−0.0001	0.0002
Education (Z2)	0.0026	0.0021	0.0010	0.0022	0.0012	0.0009
Farming experience (Z3)	−0.0017*	0.0010	0.0003	0.0010	0.0003	0.0004
Credit access (Z4)	−0.0063	0.0177	−0.0253	0.0182	0.0163**	0.0073
Payment for agro-input (Z5)	−0.0183	0.0155	0.0033	0.0162	0.0007	0.0068
Wrapping bag (Z6)	0.0051	0.0180	0.0089	0.0185	0.0061	0.0079
Market access (Z7)	−0.0115	0.0195	0.0437**	0.0194	0.0046	0.0090
Classifying sale (Z8)	0.0154	0.0165	−0.0090	0.0163	0.0036	0.0074
Land area (Z9)	0.0033***	0.0009	0.0048***	0.0009	0.0013***	0.0003

* Significant at the 10% level, ** significant at the 5% level, *** significant at the 1% level.

In the first season, the farming experience variable was negative influence on technical efficiency of DaiLoan-mango producers at the 5% significance level. This indicated that if producers rise farming experience in 10%, mango yield would decline in 0.017%. The result was against with some earlier researches [12]–[14]. The studies stated a positive relationship between technical efficiency and farming experience.

In the second season, the age variable was negative and significant effect on farmers' technical efficiency at the 10% level contrasting with being positive influence of the market access variable at the 5% probability level. The finding of age was a significant result for younger farmers were relatively more efficient than older farmers. The result was in conformity with the studies of [15]–[20]. However, the information was disagreement with some previous researches [12], [21].

In the third season, the credit access variable was positive influence on technical efficiency of DaiLoan-mango growers at the 5% significance level. The result of credit access was similar with previous studies of Kiet and Thoa [20], Bifarin et al. [22] and Khan and Ali [23] who found a strong and positive relationship between credit access and technical efficiency of the farmer, but it was different from past result of Khan and Saeed [5], Daniel [18] and Khan and Ali [23].

Particularly, the land area variable had positive coefficients and highly significant at the 1% in all three seasons at the conventional significance levels. Similar findings were obtained by Maria [12], Kiet and Thoa [20], Obare et al. [24] and Dorward [25]. However, this went against the results of Adbur [15] and Daniel [18].

3.3 Estimating the distribution of technical efficiency

The research indicated that technical efficiency was between 0.2044 and 0.8387, and a mean TE was 0.6265 in the season 1. Next, TE of the season 2 was from 0.1039 to 0.8342 with a mean of 0.5841. In season 3, TE ranged from 0.5616 to 0.8252, and achieved a mean TE 0.7310. The finding suggested TE gap of approximately 37.35%, 41.59% and 26.90% in seasons 1, 2 and 3 respectively, thereby implying that an average mango farmer in Vietnam had the capacity to rise technical efficiency in mango production by 37.35%, 41.59%, and 26.90% in seasons 1, 2, and 3 to obtain the maximum possible level.

Table 3: Frequency distribution of technical efficiency. *(Source: Field Survey Data, 2018.)*

Technical efficiency level	Season 1		Season 2		Season 3	
	Frequency	Percentage	Frequency	Percentage	Frequency	Percentage
<0.1	0	0.00	0	0.00	0	0.00
0.1–<0.2	0	0.00	1	0.40	0	0.00
0.2–<0.3	1	0.42	2	0.80	0	0.00
0.3–<0.4	8	3.35	20	8.03	0	0.00
0.4–<0.5	25	10.46	30	12.05	0	0.00
0.5–<0.6	64	26.78	78	31.33	3	1.23
0.6–<0.7	66	27.62	76	30.52	56	22.95
0.7–<0.8	69	28.87	41	16.47	169	69.26
0.8–<0.9	6	2.51	1	0.40	16	6.56
0.9–<1.0	0	0.00	0	0.00	0	0.00
1.0	0	0	0	0	0	0
Obs (N)	239		249		244	
Minimum	0.2044		0.1039		0.5616	
Maximum	0.8387		0.8342		0.8252	
Mean	0.6265		0.5841		0.7310	
Standard deviation	0.1156		0.1184		0.0518	

The sample frequency distribution showed that there were TE gap but with scope for enhancement in mango farming among mango farmers. The implication of the result was that the average mango grower required 25.30% [(1–0.6265/0.8387)*100] in season 1, 29.98% [(1–0.5841/0.8342)*100] in season 2 and 11.42% [(1–0.7310/0.8252)*100] in the second season cost saving to attain the status of the most efficient mango gardeners in Vietnam, while the least efficient gardeners proposed an enhancement in technical efficiency of 75.63% [(1–0.2044/0.8387)*100] in season 1, 87.54% [(1–0.1039/0.8342)*100] in season 2, and 31.94% [(1–0.5616/0.8252)*100] respectively.

4 CONCLUSIONS

The result of analysis presented that season 3 was the highest technical efficiency 73.10%, the second place was season 1 approximately 62.65%, and then season 2 was 58.41%. This suggested that gardeners would increase their farming on average via 73.10%, 62.65% and 58.41% respectively.

In addition, the findings indicated that adjustments in input factors could help to improved production of DaiLoan-mango in Vietnam. More specific, the independent variables that played major role in determining yield in the first season were the pesticide, root fertiliser, leaf fertiliser, and labour, in season 2 were the root fertiliser, leaf fertiliser, and labour, and in season 3 were the pesticide, root fertiliser, leaf fertiliser and labour.

Eventually, the result showed that the positive determinants of TE were the land area in all three seasons, the market access in the second season, and the credit access in the third season. However, the constraints to DaiLoan-mango production were the farming experience in the first season, and the age and in the second season.

ACKNOWLEDGEMENTS

The study used data source from project "Value chain development of Vietnamese mango fulfilling requirement for domestic and international markets" (2017–2020, code: KHCN-TNB.ĐT/14-19/C14). We would like to thank the financial support from program "Technology and science program for sustainable development in south-western region" from the Ministry of Technology and Science in Vietnam. Thanks for coordination from Professor Tran Van Hau (project leader) from school of Agriculture, Can Tho University.

REFERENCES

[1]   FAO, *Major Tropical Fruits: Statistical Compendium 2017*, FAO: Rome, 38 pp., 2019.
[2]   General Statistic Office of Vietnam (GSO), *Statistical Year Book 2017*, Statistical Publishing House, Vietnam, 2018.
[3]   Loc, V.T.T., Kiet, T.H.V.T., Son, N.P., An, N.T.T., Tin, N.H., Tho, T.H. & Huon, L., Analysis of mango value chain in Dong Thap province. Mekong Delta Development Research Institute, Can Tho University, Vietnam, 2014.
[4]   Farrell, M.J., The measurement of productive efficiency. *Journal of the Royal Statistical Society, Series A*, pp. 253–281, 1957.
[5]   Khan, H. & Saeed, H., Measurement of technical, allocative and economic efficiency of tomato farms in northern Pakistan. *International Conference on Management, Economics and Social Sciences (ICMESS 2011)*, Bangkok, 2011.
[6]   Battese, G. & Corra, G., Estimation of a production frontier model with the application of the pastoral zone of Easter Australia. *Australian Journal of Agricultural Economics*, **21**(3), pp. 167–179, 1977.
[7]   Amaza, P.S. & Olayemi, J.K., Technical efficiency in food crop production Gombe State, Nigeria, *The Nigeria Agricultural Journal,* **32**(2), pp. 140–151, 2001.

[8]   Ume, S.I., Ezeano, C.I., Chukwuigwe, O. & Gbughemobi, B.O., Resource use and technical efficiency of okra production among female headed household: Implication for poverty alleviation in the rural areas of south east, Nigeria. *International Journal of Advanced Research and Development*, **5**(24), 2018.

[9]   Aigner, D.K., Lovell, C.K. & Schmidt, P., Formulation and estimation of stochastic frontier production function models. *Journal of Econometrics*, **6**, pp. 21–37, 1977.

[10] Coelli, T.J., Prasada Rao, D.S., O'Donnell, C.J. & Battese, G.E., *An Introduction to Productivity and Efficiency Analysis*, 2nd ed., Springer: New York, USA, pp. 214–222, 2005.

[11] Ogunniyi, L.T., Profit efficiency among maize producers in Oyo State, Nigeria. *ARPN Journal of Agricultural and Biological Science*, **6**, pp. 11–17, 2011.

[12] Maria, S.M., Analysing the technical and allocative efficiency of small-scale maize farmers in Tzaneen municipality of Mopani district: A Cobb–Douglas and logistic regression approach. Master's thesis of Agricultural Management (Agricultural Economics), Department of Agricultural Economics and Animal Production, Faculty of Science and Agriculture School of Agricultural and Environmental Sciences at the University of Limpopo, 2015.

[13] Abdukadir, S., Analysis of technical efficiency of groundnut production: the case of smallholder farmers in Harari region. MSc thesis presented to the School of Graduate Studies of Haramaya University, 2010.

[14] Dadzie, S.K.N. & Dasmani, I., Gender difference and farm level efficiency: Metafrontier production function approach. *Journal of Development and Agricultural Economics*, **2**, pp. 441–451, 2010.

[15] Abdur, R.S.M., A study on economic efficiency and sustainability of wheat production in selected areas of Dinajpur District. MSc thesis, Bangladesh Agricultural University, 2012.

[16] Alam, A., Kobayashi, H., Motsumura, I., Ishida, A. & Esham, M., Technical efficiency and its determinants in potato production: Evidence from northern areas in Gilgit-Baltistan region. *International Journal of Research in Management, Economics and Commerce*, **2**, pp. 1–17, 2012.

[17] Bealu, T., Endrias, G. & Tadesse, A., Factors affecting economic efficiency in maize production: The case of Boricha Woreda in Sidama Zone, Southern Ethiopia. *The 4th Regional Conference of the Southern Nationalities State Economic Development in Hawassa*, 28 pp., 2013.

[18] Daniel, H.G., Analysis of economic efficiency in potato production: The case of smallholder farmers in Welmera district, Oromia special zone, Oromia, Ethiopia. MA thesis in Development Economics, Department of Economics, College of Business and Economics, School of Graduate Studies, Hawassa University, 2016.

[19] Sibiko, K.W., Mwangi, J.K., Gido, E.O., Ingasia, O.A. & Mutai, B.K., Allocative efficiency of smallholder common bean producers in Uganda. *International Journal of Development and Sustainability*, **2**(2), pp. 640–652, 2013.

[20] Kiet, T.H.V.T. & Thoa, N.T.K., Technical efficiency of mango in Vietnam. *International Journal of Advanced Science and Technology*, **29**(11), pp. 748–755, 2020.

[21] Malinga, N.G., Masuku, M.B. & Raufu, M.O., Comparative analysis of technical efficiencies of smallholder vegetable farmers with and without credit access in Swazil and the case of the Hhohho region. *International Journal of Sustainable Agricultural Research*, **2**(4), pp. 133–145, 2015.

[22] Bifarin, J.O., Alimi, T., Baruwa, O.I. & Ajewole, O.C., Determinants of technical, allocative and economic efficiencies in the plantain (Musa spp.) production industry, Ondo State, Nigeria. Federal College of Agriculture, Ondo State, Nigeria, 2010.
[23] Khan, H. & Ali, F., Measurement of productive efficiency of tomato growers in Peshawar, Pakistan. *Agric. Econ. Czech.*, **8**, pp. 381–388, 2013.
[24] Obare, G.A., Daniel, O.N. & Samuel, M., Are Kenyan smallholders allocatively efficient? Evidence from Irish potato producers in Nyandarua North district. *Journal of Development and Agricultural Economics,* **2**(3), pp. 78–85, 2010.
[25] Dorward, A., Farm size and productivity in Malawian smallholder agriculture. *Journal of Development Studies*, **35**, pp. 141–161, 1999.

Author index

WITPRESS ...for scientists by scientists

The Sustainable City XIV

Edited by: **G. PASSERINI**, *Marche Polytechnic University, Italy* and **S. RICCI**, *University of Rome "La Sapienza", Italy*

Urban areas result in a series of environmental challenges varying from the consumption of natural resources and the subsequent generation of waste and pollution, contributing to the development of social and economic imbalances. As cities continue to grow all over the world, these problems tend to become more acute and require the development of new solutions.

The challenge of planning sustainable contemporary cities lies in considering the dynamics of urban systems, exchange of energy and matter, and the function and maintenance of ordered structures directly or indirectly supplied and maintained by natural systems. The task of researchers, aware of the complexity of the contemporary city, is to improve the capacity to manage human activities, pursuing welfare and prosperity in the urban environment.

Any investigation or planning for a city ought to consider the relationships between the parts and their connections with the living world. The dynamics of its networks (flows of energy-matter, people, goods, information and other resources) are fundamental for an understanding of the evolving nature of today's cities.

Large cities are probably the most complex mechanisms to manage. They represent a fertile ground for architects, engineers, city planners, social and political scientists, and other professionals able to conceive new ideas and time them according to technological advances and human requirements.

Papers presented at the 14th International Conference on Urban Regeneration and Sustainability address the multidisciplinary components of urban planning, the challenges presented by the increasing size of cities, the number of resources required and the complexity of modern society. Various aspects of the urban environment are covered and a focus is placed on providing solutions which lead towards sustainability.

ISBN: 978-1-78466-413-8   eISBN: 978-1-78466-414-5
Published 2020 / 190pp

www.ingramcontent.com/pod-product-compliance
Lightning Source LLC
Chambersburg PA
CBHW062008190326
41458CB00009B/3011